Conceptual Chemistry

Understanding Our World of Atoms and Molecules

Third Edition

Laboratory Manual

Donna Gibson

Chabot College

John Suchocki

Saint Michael's College

San Francisco • Boston • New York
Capetown • Hong Kong • London • Madrid • Mexico City
Montreal • Munich • Paris • Singapore • Sydney • Tokyo • Toronto

Publisher:	Jim Smith
Project Editor:	Katherine Brayton
Executive Marketing Manager:	Scott Dustan
Managing Editor:	Corinne Benson
Production Supervisor:	Mary O'Connell
Production Management and Composition:	Simmy Cover and Craig Johnson
Cover Image:	Steve Simonsen
Cover Design:	17th Street Studios
Copyeditor:	Irene Nunes
Artist:	Shirley Bortoli
Manufacturing Buyer:	Pam Augspurger

ISBN: 0-8053-8232-1

6 7 8 9 10 —TCS— 10 09

www.aw-bc.com

Contents

Preface v

1 Laboratory Safety and Common Techniques 1

2 Taking Measurements 11

3 Physical and Chemical Properties and Changes 19

4 Percent Water in Popcorn 27

5 Salt and Sand 33

6 Radioactivity 37

7 Bright Lights 45

8 Electron-Dot Structures 51

9 Molecular Shapes 57

10 Solutions 63

11 Candy Chromatography 73

12 How Much Fat? 79

13 Energy and Calorimetry 83

14 The Clock Reaction 91

15 Upset Stomach 97

16 Mystery Powders 105

17 Electrochemistry 111

18 Organic Molecules 119

19 DNA Capture 129

Appendix 133

Preface

To the Student

Learning the concepts of chemistry without experiencing them for yourself in the laboratory is akin to looking through a travel brochure without ever traveling. To get the full experience, you've got to get up and see things for yourself.

Consider this laboratory manual as a road map. Follow its directions and you will find yourself driving by some of the main attractions of chemistry. On occasion, however, you may find it necessary to veer off the main highway to get to where you need to be. In other words, you may find that the procedures given or the equipment made available to you are insufficient thus requiring you to innovate. This is okay, and actually desirable, so long as you do so under the careful guidance of your instructor.

The wise traveler plans his or her trip before departing. You'll be doing yourself and the rest of your class a big favor, therefore, by taking the time to look over an assigned activity before coming to lab.

Also, you aren't traveling alone. Please be sure to take advantage of this by articulating your observations and thought processes with your fellow classmates and your tourguide instructor. The more you talk about your experiences, the better they will "sink in."

Have a safe and pleasant journey through this enjoyable world of chemistry!

To the Instructor

Thank you for selecting this laboratory manual for your liberal arts chemistry course. Our aim has been to provide laboratory activities that are safe, interesting, and also of a level that corresponds to the companion textbook *Conceptual Chemistry.* Please refer to the *Conceptual Chemistry Instructor's Manual* for our detailed discussions about these laboratories. Included you will find instructions for setting up various activities, alternate procedures, answers to the questions, and more.

Much effort has been put forth to keep this manual as error-free and accurate as possible. In all likelihood, however, some errors and/or inaccuracies will have escaped our notice. Your help in forwarding to us any errors or inaccuracies that you catch would be greatly appreciated. Your questions, general comments, and criticisms are also most welcome. We can be contacted through Support@ConceptChem.com. We look forward to hearing from you.

Donna Gibson, Chabot College

John Suchocki, St. Michael's College

Acknowledgements

We owe great thanks to the team at Benjamin Cummings for their patience, encouragement, and expertise, particularly, Jim Smith, Lisa Leung, and Tony Asaro. For development of the manuscript we thank Jean Lake and Irene Nunes. Special thanks are also extended to Mary Graff of Amarillo College and to Donna's colleagues at Chabot College for their contributions and advice, in particular, Laurie O'Connor, Maggie Schumacher, Sue Stanton and Vince Triggs. We are also most grateful to the following reviewers for their many helpful suggestions: Ana Gaillat, Greenfield Community College; Shelley Gaudia, Lane Community College; Kevin Johnson, Pacific University; Art Maret, University of Central Florida; and Frank Palocsay, James Madison University.

We dedicate this book to Donna's husband, Pete Gibson, whose moral support and strong family commitment made this project possible. A special recognition is also extended to Donna's children, Sierra, Zach, and Alex, for their love and understanding through the duration of this project.

1 Laboratory Safety and Common Techniques

Objectives

- To become familiar with basic safety rules in a chemistry laboratory
- To become familiar with the location and use of basic safety equipment in the laboratory
- To become familiar with glassware and basic laboratory techniques

Materials Needed

Equipment
- filter paper
- glass funnel
- funnel support
- four 250-mL beakers
- glass stirring rod
- Buchner funnel
- neoprene adapter
- 250-mL filter flask
- rubber tubing
- water-trap bottle fitted with two-hole stopper
- two short lengths of glass tubing
- centrifuge
- two centrifuge tubes
- Bunsen burner
- matches
- electronic balance
- 100-mL graduated cylinder

Chemicals
- white distilled vinegar
- milk

Discussion

Most safety rules that must be followed in the chemistry laboratory involve common sense. If you are ever unsure of a procedure or chemical, *ask your laboratory instructor,* who is there to help you.

As with any other place, accidents can occur, and therefore it is important to know the location and proper use of the safety equipment in your laboratory. Learning that information is therefore part of this first laboratory session. The remainder of the session introduces you to some frequently used equipment and techniques.

Part A Safety

A blanket instruction for this entire course is that

> **You must wear safety goggles at all times in the chemistry laboratory, from the moment you unlock your drawer to the moment you sign out.**

There are no mentions in individual experiments about wearing safety goggles because you are expected to have them on *at all times.* Repeat—at all times. That being said, here are some safety guidelines you should always follow.

Laboratory Safety Rules—Self-Protection

Wear safety goggles at all times.

Wear closed-toed shoes, not sandals.

Tie back long hair to prevent it from coming in contact with chemicals or flames.

Report any injury to your laboratory instructor immediately.

Laboratory Safety Rules—Procedures

1. Do not eat, drink, or smoke in the laboratory.
2. Maintain a clean and orderly work space. Clean up spills at once or ask for assistance in doing so.
3. Do not perform unauthorized experiments.
4. Do not taste any chemicals or directly breathe any chemical vapors.
5. Check all chemical labels for both name and concentration.
6. Do not grasp recently heated glassware, clamps, or ring stands because they remain hot for quite a while.
7. Discard all excess reagents or products in the proper waste containers. **Important:** Most chemicals cannot be poured down the sink.
8. If your skin comes in contact with a chemical, rinse under cold water for at least 15 minutes.
9. Do not work with flammable solvents near an open flame.
10. Assume any chemical is hazardous if you are unsure.

Safety Equipment in the Laboratory

The following safety equipment should be present in your laboratory. Your instructor will point out the location of each and demonstrate its use. On the report sheet, printed at the end of this chapter, make a sketch of the laboratory and identify their locations.

1. Eyewash station
2. Fire extinguishers
3. Fire blankets
4. Safety showers
5. Fume hoods
6. Main gas shutoff valve

Part B Laboratory Glassware and Equipment

Figure 1 shows most of the glassware and equipment you'll be using in the weeks ahead.

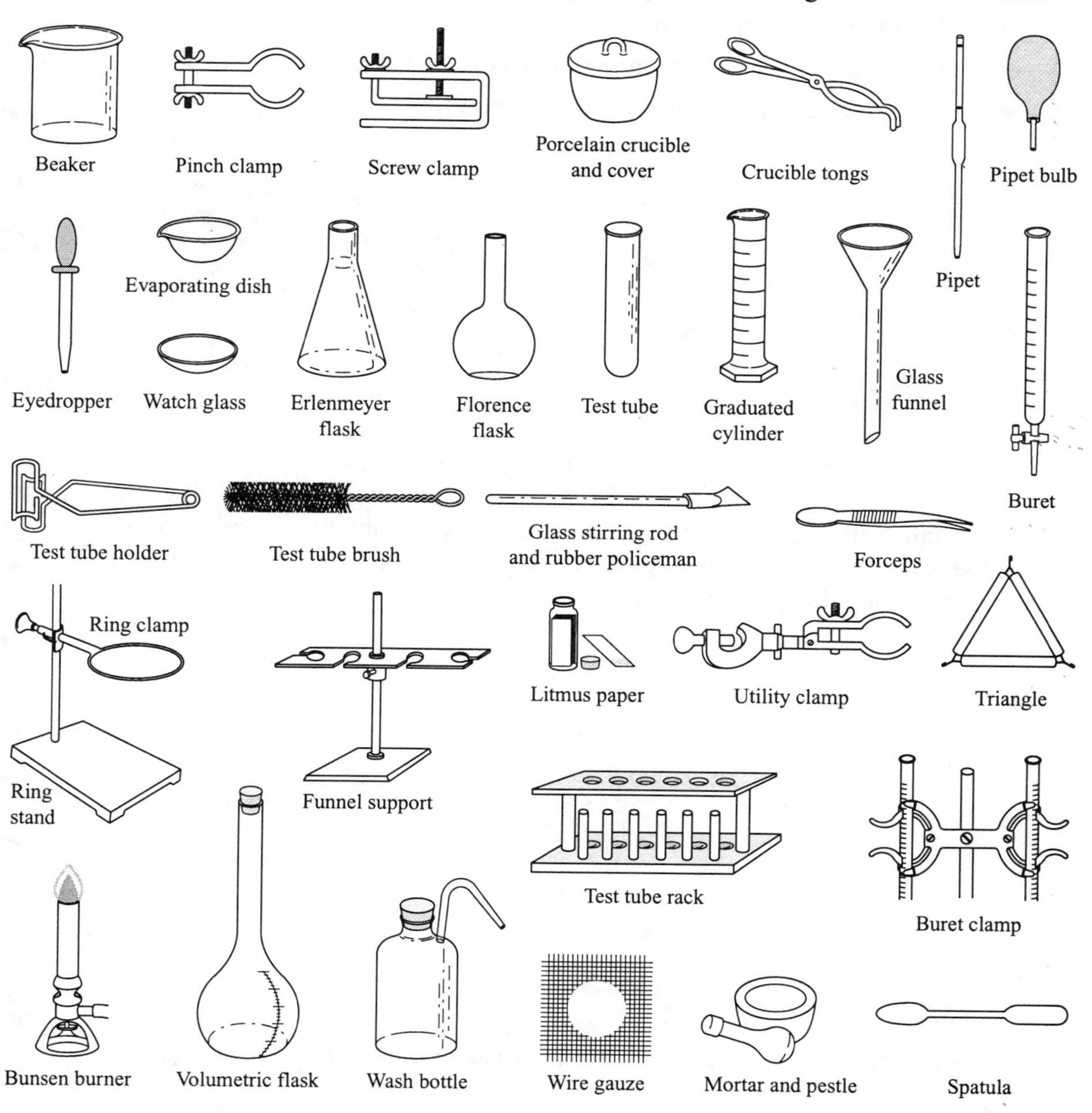

FIGURE 1

Part C Common Laboratory Techniques

1. Gravity Filtration

Filtration is a technique commonly used to separate a solid from a liquid. The liquid can either be pure (distilled water, for instance) or a solution (such as salt water). The solid–liquid mixture is poured through a filter paper that allows only the liquid to pass through its pores. The liquid that goes through the filter paper is referred to as the filtrate, and the solid remaining on the filter paper is referred to as the precipitate.

Figure 2 shows the correct way to fold a filter paper:

1. Fold the filter paper in half.
2. Fold again, but leave the top quarter a little short.
3. Tear off a little corner of this short quarter and discard.
4. Open up the larger quarter section to form a cone and insert into a funnel.
5. Moisten with water and seal against the funnel wall.

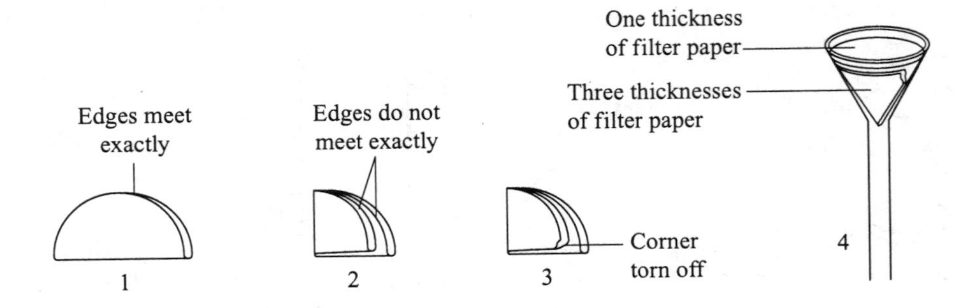

FIGURE 2 Steps in folding a filter paper.

The mixture to be filtered must be poured directly into the filter-paper cone and not down the side of the funnel. Use a glass stirring rod as a pouring aid to direct the flow and to prevent splashing, as shown in Figure 3.

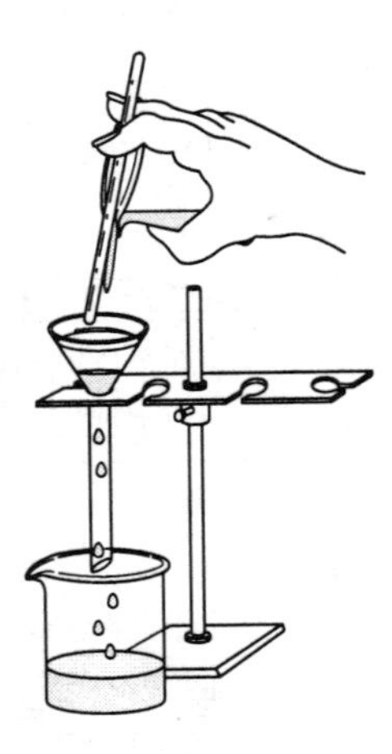

FIGURE 3 Technique for pouring a liquid-solid mixture into a funnel fitted with filter paper.

2. Vacuum Filtration

Vacuum filtration requires the specialized equipment shown in Figure 4. A water aspirator or vacuum pump is required to create a vacuum. This type of filtration is more rapid than gravity filtration. A water trap is commonly used to prevent water from backing up into the system as the water pressure through the aspirator changes.

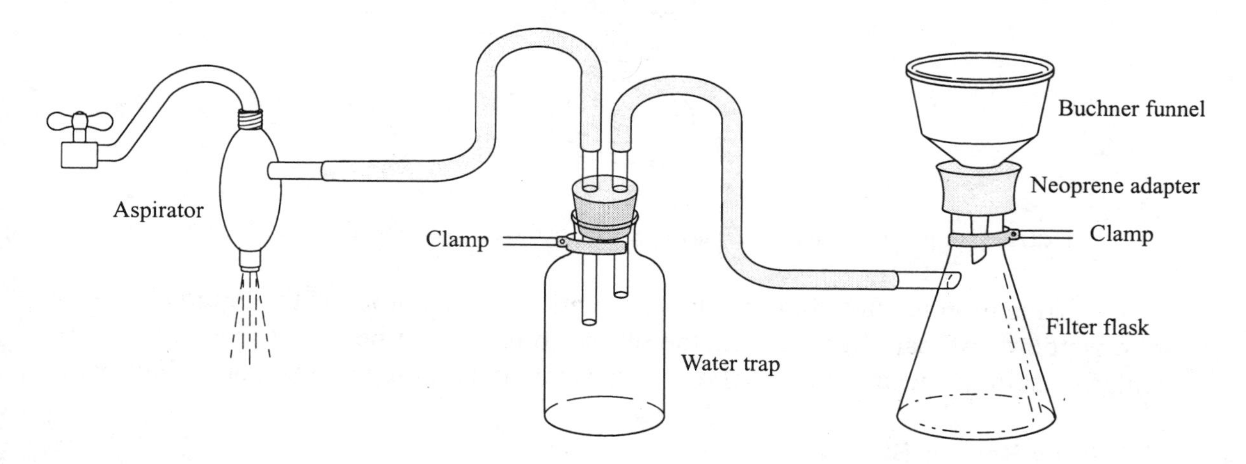

FIGURE 4 A typical setup used in vacuum filtration.

When performing a vacuum filtration, be sure that the rubber hoses are connected tightly and that the Buchner funnel fits snugly into the neoprene adapter. Before pouring your mixture into the funnel, test the suction by turning on the water that will run through the aspirator and placing the palm of your hand over the Buchner funnel to feel the suction. Then place a piece of filter paper flat in the funnel and wet it slightly with the solvent being used. After you have poured all your mixture in and filtration is complete, disconnect the rubber hose from the filter flask before turning off the water running through the aspirator.

3. Centrifugation

A centrifuge causes the solid and liquid components of a mixture to separate by spinning the mixture at a very high speed. Whenever you use a centrifuge, it must be balanced while it is spinning. Therefore, if you have only one tube to centrifuge, prepare another tube filled with water to the same level. Insert the two tubes on opposite sides of the centrifuge, as shown in Figure 5. Close the lid, and turn on the power switch. One minute of centrifuging at full speed is usually sufficient. Do not open the lid until the rotor has stopped. Keep your hands away from the top of the centrifuge while it is rotating.

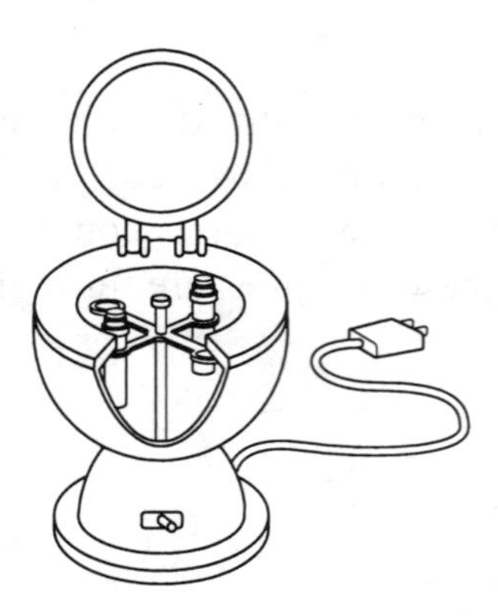

FIGURE 5 A properly balanced centrifuge, with two tubes placed opposite each other.

Centrifugation is often followed by decantation. In this process, the liquid above the solid is poured off carefully, leaving the settled solid undisturbed. This technique is particularly useful if the amount of solid is very small and might get "lost" on a filter paper.

4. Lighting a Bunsen Burner

A Bunsen burner is commonly used as a heat source in the chemistry laboratory. Figure 6 shows a typical unit.

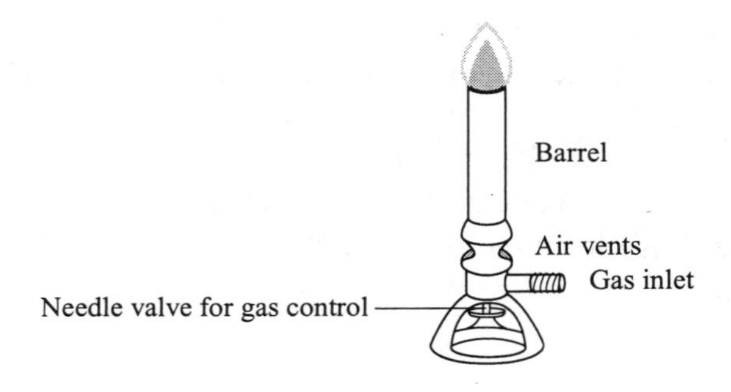

FIGURE 6 A typical Bunsen burner.

Here is the correct procedure for lighting a Bunsen burner:

1. Close the gas inlet valve completely, and then carefully open it about two complete turns. Doing this prevents the flame from being very high when the burner is first lit.
2. Turn on the gas source and hold a lit match to the top of the barrel.
3. Adjust the air vents until a double cone of flame appears. When the flame is adjusted properly, you will see a dark blue cone within an outer light blue cone. This is shown in Figure 7.

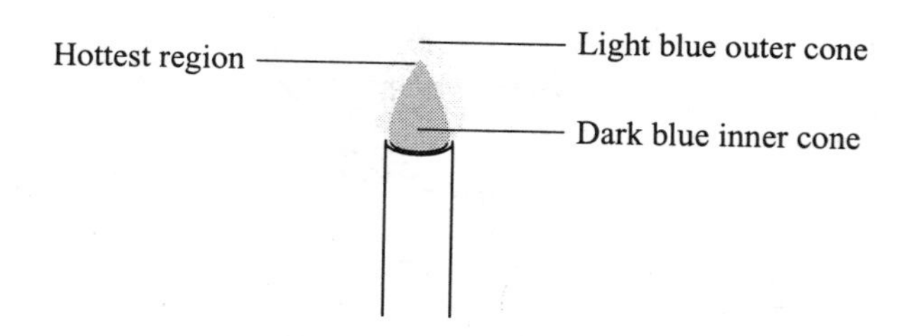

FIGURE 7 A properly adjusted flame on a Bunsen burner.

5. Heating with a Bunsen Burner

Figure 8 shows for the proper way to heat substances either in a test tube or in a beaker or flask. Note that the test tube should be no more than half full. Move the flame around the whole bottom half of the test tube, rather than focusing on just one spot. Use wrist action to keep the contents in a state of gentle agitation. Be sure the mouth of the tube is not pointed at anybody, including yourself.

You may often use a Bunsen burner to heat a beaker of liquid sitting on a ring clamp attached to a ring stand. Adjust the height of the ring clamp so that the tip of the inner cone of the flame is just touching the wire gauze holding the beaker as shown in Figure 8.

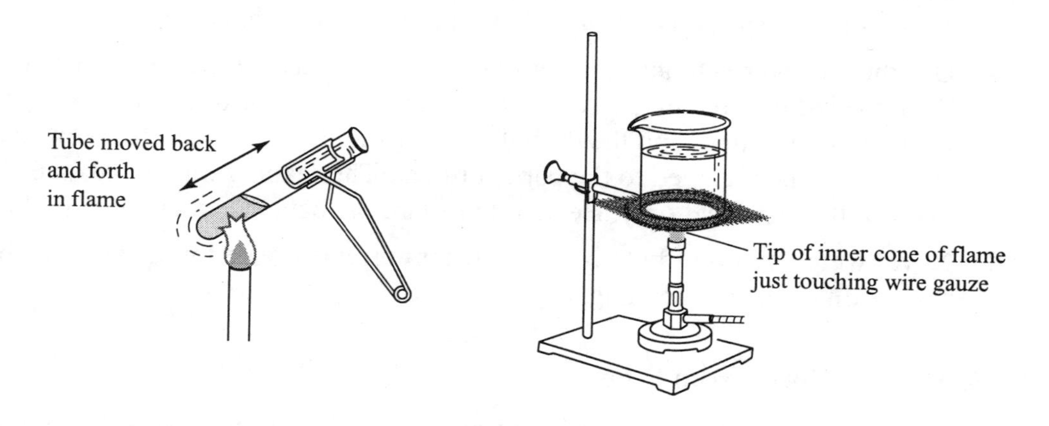

FIGURE 8 Proper heating technique using a Bunsen burner.

6. Using an Electronic Balance

A typical electronic balance is shown in Figure 9. These devices are very expensive and should be used only after your instructor has demonstrated the correct way to use the particular model in your laboratory.

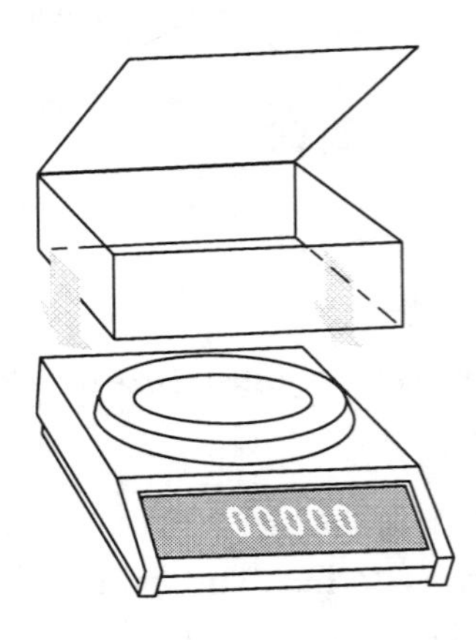

FIGURE 9 An electronic balance.

Here are some common practices when using an electronic balance:

1. Never place any chemical directly on the balance. Chemicals should be placed either on a piece of weigh paper or in a weigh boat.
2. Be sure the balance reads zero before you place anything on the weigh pan. If the reading is not zero, push the tare button to re-zero the balance.
3. Use the tare button to automatically subtract the mass of the weigh paper or boat. With the balance reading zero, place the empty paper or boat on the weigh pan. Push the tare button, and the digital readout should return to zero. Now add the chemical to be weighed to the paper or boat. The reading on the balance is the mass of the chemical only. Be sure to re-tare the balance when you are done.
4. Never weigh warm or hot objects. The air currents created will cause the balance to give an erroneous reading.

Part D Separating Curds from Whey

Now that you have learned about several laboratory techniques, the task at hand is to determine the mass of the curds contained in 50 mL of milk. To precipitate the curds, pour 50 mL of the milk assigned to you by your instructor into a 150-mL beaker. Then slowly stir 15 mL of white distilled vinegar into the milk. This milk may be whole milk, 2% milkfat, 1% milkfat, or skim milk. Make a guess as to which technique—gravity filtration, vacuum filtration, or centrifugation—will separate the curds most effectively. Try your method and measure the mass of the curds you collect. Make sure your sample is as dry as possible before weighing. Compare your results with those of your classmates.

Name ____________________________

Laboratory Safety and Common Techniques Report Sheet

Part A Safety

Sketch of laboratory showing location of all safety equipment:

Part D Separating Curds from Whey

1. Write out the procedure you followed to separate the curds. Be sure to include equipment, quantities of chemicals, and a step-by-step outline of what you did.

2. Which type of milk did you work with, and what mass of curds were you able to isolate? Compare your results with those of your classmates and draw a conclusion.

Type of milk (circle one): Whole 2% milkfat 1% milkfat skim milk

Mass of curds isolated: ____________ g

2 Taking Measurements

Conceptual Chemistry Laboratory

Objectives

- To read a measurement scale to the correct number of significant figures
- To distinguish between accuracy and precision
- To determine the density of a liquid
- To identify an unknown liquid by determining its density
- To determine the density of a regular solid and an irregular solid

Materials Needed

- 10-mL graduated cylinder
- 100-mL graduated cylinder
- 50-mL beaker
- eyedropper
- balance
- metric ruler
- samples of known and unknown liquids
- samples of regular and irregular solids

Discussion

Much of what is done in the chemistry laboratory involves taking measurements. A measurement is a quantitative observation that has both a numerical value and a unit.

How well a measurement is taken determines both its precision and its accuracy. *Precision* is related to the reproducibility of the measurement. It is a comparison of several measured values obtained in the same way. For example, a student measures the volume of a liquid three times, obtaining the values 3.66 mL, 3.62 mL, and 3.64 mL. These measurements have high precision; there is only a 0.040-mL difference between the highest and lowest values.

Accuracy is a comparison of a measured value to the accepted, or true, value. Suppose the volume of the liquid had actually been 5.44 mL. Then the accuracy of the three measurements taken would be low because a difference of 1.80 mL between the average measured value and the true value is relatively large.

A good scientist is always trying to achieve both high precision and high accuracy in the laboratory.

Measuring Length

All measurements are estimated values. They are obtained by using a measuring device that is marked with a calibrated scale. The scale contains marks and printed numeric values at regular intervals. There is a degree of uncertainty in all measured values. The degree of uncertainty is indicated by the number of significant figures given in a measurement (see Appendix). For example, of the two rulers shown in Figure 1, is one more uncertain than the other?

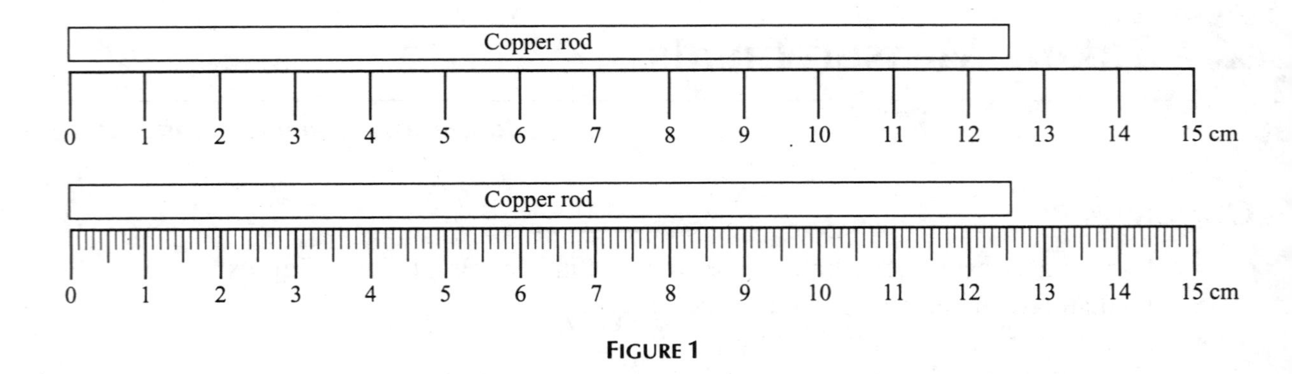

FIGURE 1

In the top ruler, each graduation represents 1 cm. It is certain that the right end of the copper rod lies between the 12th and 13th graduations. By estimating only one digit, the reading can be taken as 12.5 cm. In the bottom ruler, the larger graduations again represent 1 cm, and again the end of the rod lies between the 12th and 13th graduations. This ruler also has smaller graduations, however, representing 0.1 cm, and the rod lies between the 5th and 6th smaller graduations. By estimating only one digit, the reading can be taken as 12.55 cm. We see, therefore, that the ruler with many smaller graduations allows for greater precision. Whereas the top ruler allows measurement of around ±0.5 cm, the bottom ruler with more graduations allows measurements of around ±0.05 cm.

Measuring Volume

When taking volume measurements, you will notice that the surface of the liquid is curved. This curved surface is called a *meniscus*. You should always take your reading from eye level and report the position of the *bottom* of the meniscus (See Figure 2).

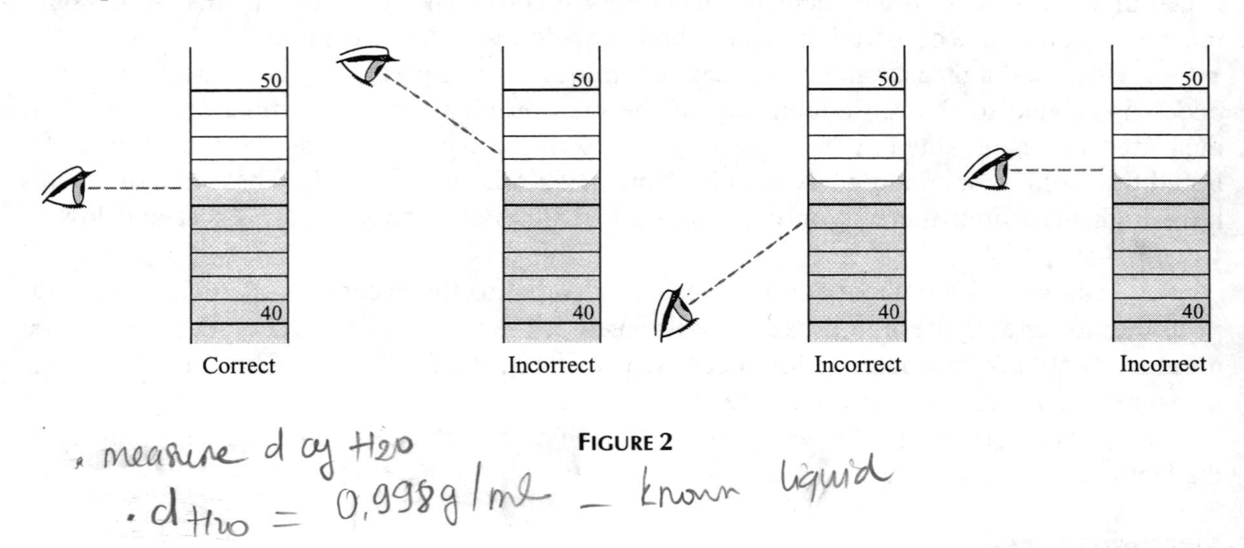

FIGURE 2

• measure d of H_2O
• d_{H_2O} = 0.998 g/ml – known liquid

Measuring Density

In this laboratory, you will practice taking measurements and recording them to the correct degree of uncertainty while determining the density of various samples. Density is defined as mass per unit volume and can be calculated using the equation

$$\text{density} = \frac{\text{mass of sample}}{\text{volume of sample}}$$

Once you determine a density experimentally, you will need to evaluate the precision and accuracy of your results. We shall use a calculation called *percent range* to express precision and a calculation called *percent error* to express accuracy. The necessary formulas are

$$\text{percent range} = \frac{\text{highest experimental value} - \text{lowest experimental value}}{\text{average experimental value}} \times 100$$

$$\text{percent error} = \frac{\text{true value} - \text{average experimental value}}{\text{true value}} \times 100$$

Procedure

Part A Volume Measurements

1. Add approximately 8 mL of water to a 10-mL graduated cylinder.
2. Read the water volume from the marks on the cylinder and record your reading on the report sheet, using the correct number of significant figures.
3. Carefully pour this water into a 100-mL graduated cylinder.
4. Read the water volume from the marks on the cylinder and record your reading on the report sheet, using the correct number of significant figures.
5. Carefully pour all the water into a 50-mL beaker.
6. Read and record the water volume, using the correct number of significant figures.

Part B Density of Liquids

Known Liquid Sample

1. Determine and record the mass of an empty 10-mL graduated cylinder.
2. Add approximately 5 mL of a known liquid to the cylinder.
3. Determine and record the mass of the cylinder plus the liquid.
4. Calculate the mass of the liquid.
5. Read and record the liquid volume from the marks on the cylinder, using the correct number of significant figures.
6. Empty and dry the cylinder (use the appropriate waste container).
7. Repeat the procedure, this time using approximately 8 mL of the liquid.
8. Repeat the procedure, this time using approximately 10 mL of the liquid.

Unknown Liquid Sample

1. Empty and dry the graduated cylinder.
2. Record the number of your unknown liquid.
3. Repeat the above steps using the unknown liquid.
4. To identify your unknown liquid, compare your results to those in the table provided to you by your instructor.

Part C Density of Solids

Regular Solid

1. Record your sample number.
2. Determine and record the mass of your sample. (**Note:** sample must be either cylindrical or rectangular.)
3. If your sample is rectangular, measure its length, width, and height. If your sample is cylindrical, measure its diameter and height.
4. Using the formula $V = L \times W \times H$ for a rectangular solid or the formula $V = \pi r^2 \times H$ for a cylindrical solid, determine the volume of your sample.
5. Calculate the density of your sample.

Irregular Solid

1. Record your sample number.
2. Determine and record the mass of your sample.
3. Add approximately 50 mL of water to a 100-mL graduated cylinder.
4. Read and record the water volume.
5. Carefully place your sample in the cylinder.
6. Read and record the new water volume.
7. Calculate the volume of your sample.
8. Calculate the density of your sample.

3 Physical and Chemical Properties and Changes

Objectives

- To identify various physical and chemical properties of matter
- To distinguish between chemical changes and physical changes

Materials Needed

Equipment

- thermometer with cork holder
- ring stand with two clamps
- hot plate
- 250-mL and 500-mL beakers
- large test tube
- boiling chips
- glass stirring rod
- well-plates (ceramic or glass)
- eyedroppers
- microspatula
- evaporating dishes
- ice
- medium test tube with stopper

Chemicals

- various elements and compounds
- methanol
- iodine crystals
- sucrose crystals
- acetone
- steel wool
- cupric sulfate pentahydrate crystals
- 10% sodium carbonate solution
- 10% sodium sulfate solution
- 6 *M* HCl
- 10% sodium chloride solution
- 10% calcium chloride solution
- luminol crystals
- sodium perborate crystals

Discussion

Chemistry is the study of matter. It is very common for a chemist to need to describe a bit of matter as thoroughly as possible. In doing so, the chemist would certainly list *physical properties.* Many physical properties can be observed using our senses; color, crystal shape, and phase at room temperature are some examples. Other physical properties involve quantitative observations and so must be measured; density, specific heat capacity, and boiling point are three examples. A *physical change* is any change in a substance that does not involve a change in its chemical composition. During a physical change, no new chemical bonds are formed, and so the chemical composition remains the same. Examples of physical change are boiling, freezing, expanding, and dissolving.

Matter can also be characterized by its *chemical properties.* The chemical properties of a substance include all the chemical changes possible for that substance. A *chemical change* is one in which the substance is transformed to a new substance. That is, there is a change in the chemical composition of the substance. During a chemical change, the atoms are pulled apart from one another, rearranged, and put back in a new arrangement. Examples of chemical change are burning, rusting, fermenting, and decomposing.

In this experiment, you will first identify and record various physical properties of substances, using qualitative observations, such as changes in color or phase, and quantitative observations, such as boiling points. In the second part, you will look at changes in matter and determine if they are physical or chemical.

Procedure

Part A Physical Properties

1. Examine the various substances provided by your instructor and record your observations in Table 1 of the report sheet (**Note:** some substances may be toxic. As a precaution, do not open any containers without the permission of your instructor.)
2. Assemble the apparatus shown in Figure 1. Add about 300 mL of water to the 500-mL beaker and about 3 mL of methanol, CH_3OH, to the test tube. Do not forget to add the boiling chips to the test tube. Turn on the hot plate to medium. Use a stirring rod to stir the water while it is being heated, and pay close attention to the thermometer readings. Note that as the methanol boils, its temperature remains constant. Record the boiling point.

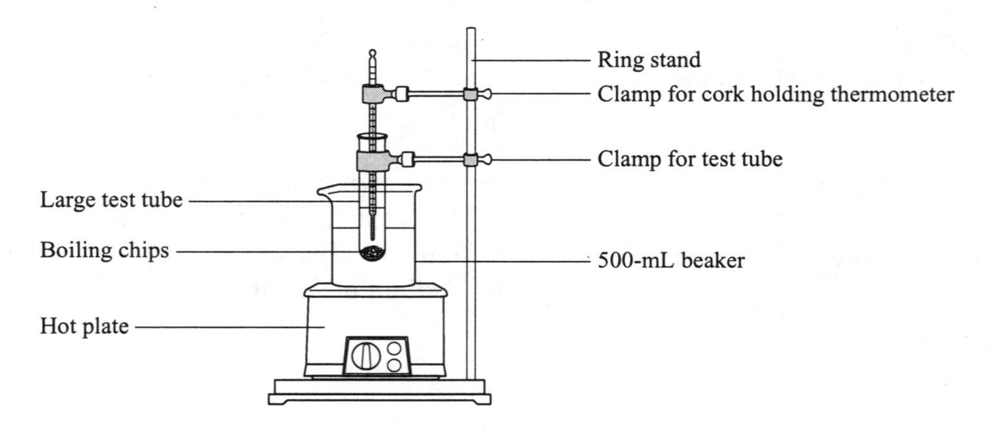

FIGURE 1

3a. Place a small crystal of iodine in a well of one well-plate and a small crystal of sucrose in a well of a second well-plate. Use an eyedropper to fill each well with distilled water and stir gently with a microspatula. Record whether each substance is completely soluble, partially soluble, or insoluble. Rinse the iodine into a designated waste container and the sucrose into the sink.

3b. Repeat the procedure using acetone as the solvent. You may need to rinse the iodine into another designated waste container (ask your instructor). The sucrose can be rinsed into the sink with water.

Part B Physical and Chemical Changes

Complete Table 2 of the report sheet for each of the following systems.

1. Inspect a small piece of steel wool. Place it in an evaporating dish, and heat on a hot plate set to high. Allow the system to cool to room temperature. Observe and record any changes in the steel wool.
2. Inspect some cupric sulfate pentahydrate crystals, $CuSO_4 \cdot 5H_2O$. Place a few crystals in an evaporating dish, and heat on a hot plate set to medium. Observe and record any changes in the salt. After the system has cooled to room temperature, add a few drops of water to the crystals. Observe and record any changes.
3. Place a few drops of a 10% sodium carbonate solution, Na_2CO_3, in one well of a well-plate and a few drops of a 10% sodium sulfate solution, Na_2SO_4, in a second well of the same well-plate. Add 2 or 3 drops of 6 *M* hydrochloric acid to each well. Observe and record any changes.
4. Place a few drops of a 10% sodium chloride solution, NaCl, in one well of a well-plate and a few drops of a 10% calcium chloride solution, $CaCl_2$, into a second well of the same well-plate. Add several drops of a 10% sodium carbonate solution to each well. Observe and record any changes.
5. Inspect some iodine crystals, I_2. Place a few of the crystals in a dry 250-mL beaker and cover with an evaporating dish that contains ice, as shown in Figure 2. In a fume hood, place the beaker on a hot plate set to medium. Observe and record any changes.
6. Fill a medium test tube that can be stoppered about halfway with distilled water. Using a microspatula, add one scoop of luminol crystals and one scoop of sodium perborate crystals. Then add one small crystal of cupric sulfate pentahydrate. Quickly stopper the test tube, and take it to a darkened room. Observe and record any changes.

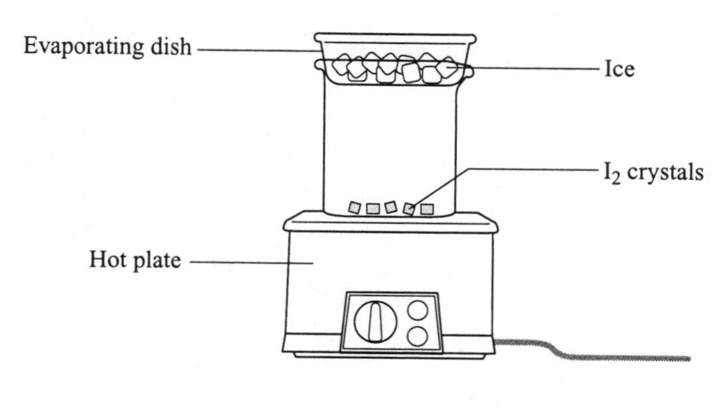

FIGURE 2

Name ____________________________

Physical and Chemical Properties and Changes Report Sheet

Part A Physical Properties

1. Complete Table 1.

TABLE 1

Name of Substance	Chemical Formula	Phase at Room Temperature	Color	Other Physical Properties Observed	Element or Compound?

2. Boiling point of methanol __________ °C
3. Solubility

   Iodine in water __________ Sucrose in water __________

   Iodine in acetone __________ Sucrose in acetone __________

Part B Physical and Chemical Changes

1. Complete Table 2.

TABLE 2

Procedure	Observation	Physical Change or Chemical Change?	Evidence or Reasoning
1. steel wool + heat			
2a. $CuSO_4 \cdot 5H_2O$ + heat			
2b. $CuSO_4 + H_2O$			
3a. Na_2CO_3 + HCl			
3b. Na_2SO_4 + HCl			
4a. NaCl + Na_2CO_3			
4b. $CaCl_2$ +Na_2CO_3			
5. I_2 + heat			
6. Luminol + sodium perborate + $CuSO_4$			

Name ______________________________

Questions

1. Distinguish between a qualitative observation and a quantitative one. Give an example of each from this experiment.

2. Classify the following properties of sodium metal as physical or chemical:
   a. silver metallic color ____________________
   b. turns gray in air ____________________
   c. melts at 98°C ____________________
   d. reacts explosively with chlorine ____________________

3. Classify the following changes as physical or chemical:
   a. steam condenses to liquid water on a cool surface ____________________
   b. baking soda dissolves in vinegar, producing bubbles ____________________
   c. mothballs gradually disappear at room temperature ____________________
   d. baking soda loses mass as it is heated ____________________

4 Percent Water in Popcorn

Conceptual Chemistry Laboratory

Objectives

- To work with the concept of percent
- To take mass measurements using a balance
- To use a Bunsen burner
- To determine the mass percent of water in popcorn

Materials Needed

- popcorn kernels (various brands)
- ring stand with ring clamp
- 250-mL beaker
- evaporating dish that fits on top of beaker
- watch glass
- Bunsen burner
- balance
- wire gauze

Discussion

Popcorn kernels are corn seeds consisting of a variety of edible compounds, mostly starch encased in a hard shell. Also encased in the shell is a small amount of liquid water. When heated, this liquid water changes to water vapor that expands and creates pressure inside the kernel. This pressure builds to the point where the hard shell explodes. As the water vapor rapidly escapes, it causes the starchy material to fluff up, and the popcorn has "popped."

In this experiment, you will determine what percentage of the total mass of a sample of popcorn kernels is water. You will do this by measuring the masses of the kernels before and after they have been cooked. The amount of water lost is assumed to be about equal to the amount of water originally contained in the kernels, a quantity that you will calculate and report in terms of mass percent.

Mass Percent

Like any other kind of percent, mass percent is a way of identifying how much of some part there is in a whole. Always based on a scale of 100, the general formula for any percent is

$$\text{percent (\%)} = \frac{\text{part}}{\text{whole}} \times 100$$

For example, if 35 people have brown eyes in a group of 70 people, the percent of people having brown eyes is

$$\text{percent with brown eyes} = \frac{35 \text{ people}}{70 \text{ people}} \times 100 = 50\%$$

The *mass percent* (or *percent by mass*) of a given component contained in a sample is 100 multiplied by the ratio of the mass of that component to the mass of the entire sample. To determine the mass percent of water in popcorn kernels, you divide the mass of the water by the total mass of the kernels and then multiply the quotient by 100:

$$\text{mass percent of water in kernels} = \frac{\text{mass of water in kernels}}{\text{total mass of kernels}} \times 100$$

In this experiment, it does not matter if your sample contains one or more "duds" (kernels that do not pop). Duds are usually kernels that were cracked during processing or shipping. When these kernels are heated, the water vapor escapes slowly through the cracks so that the pressure inside is never strong enough to explode the shells. The duds still lose almost all their water, however, and so they should not be discounted when you do your calculations.

Procedure

1. Set up the apparatus as shown in Figure 1. Note that you are basically making your own hot-air popper. The Bunsen burner provides the heat (a hot plate can be used instead), the beaker holds the air that will be heated and ultimately pop the popcorn, the evaporating dish holds the kernels, and the watch glass acts as a lid.
2. Using a balance as described by your instructor, determine the mass of the empty evaporating dish and record the mass on your report sheet.
3. Add about ten kernels to the evaporating dish and then determine and record the combined mass of the dish plus kernels.
4. Place the evaporating dish on the beaker and cover the dish with a watch glass.
5. Turn on the Bunsen burner using the procedure, given in Laboratory 1, Laboratory Safety and Common Techniques
6. Adjust the height of your ring stand so that the tip of the inner cone is touching the wire gauze holding the beaker. (Remove glassware or hold onto it carefully while adjusting the height.)
7. Heat the kernels until they pop. If all the kernels do not pop, continue heating until the popped corn start to turn brown, but avoid charring your sample.
8. After the popping is done, turn off the gas.
9. Allow the system to cool to room temperature. **Warning:** it stays hot for a few minutes, and so do not burn yourself by being impatient. Also, if you place a hot sample on a sensitive balance, convection currents in the air will throw off the balance, giving you inaccurate readings.

10. Determine and record the mass of the evaporating dish containing your popped corn (plus any duds).

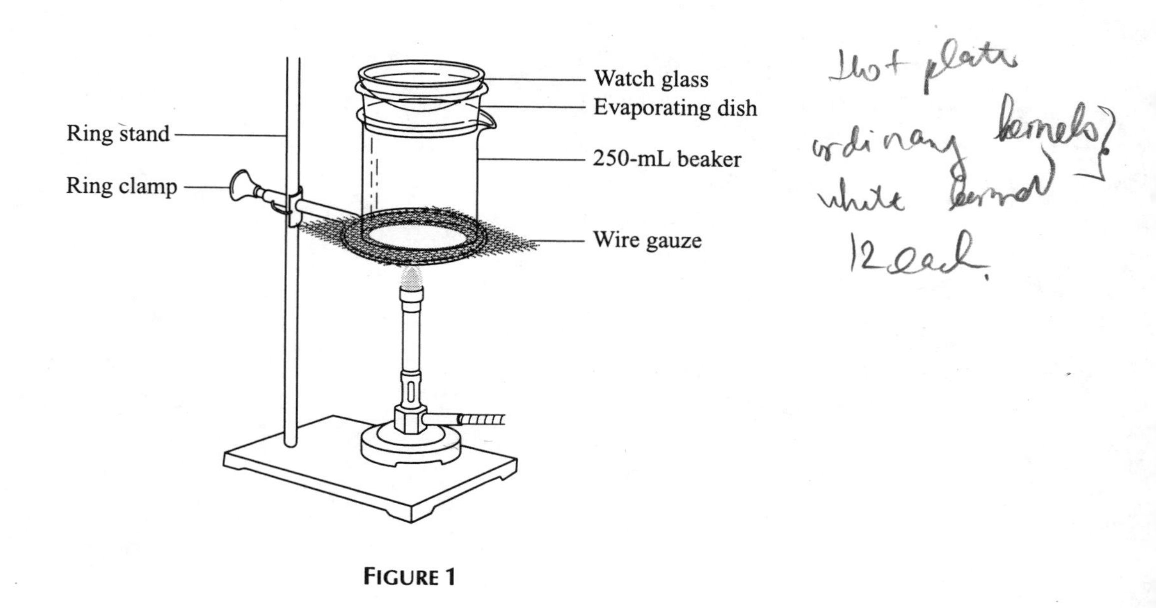

FIGURE 1

11. Repeat the procedure using a different brand of popcorn.

5 Salt and Sand

Objectives

- To develop a laboratory procedure for separating the components in a mixture of salt and sand
- To carry out the separation
- To calculate the mass percent of salt and of sand in the mixture
- To calculate the percent error in your results

Materials Needed

- various salt–sand mixtures
- various pieces of laboratory equipment, depending on the procedure you develop

Discussion

Materials can be classified as being either pure substances or mixtures. Most of the materials we encounter in our daily lives are mixtures. A mixture is a material that can be separated into two or more pure substances by physical processes.

In this experiment, you will use the properties of the components of a mixture of salt and sand to effectively separate the components from each other. Remember that physical properties describe physical attributes of a substance. They include color, density, solubility, and phase. During a physical change, one or more of the physical attributes transforms, but the chemical identity of the substance does not change. Chemical properties characterize the ability of a substance to transform to a different substance. During a chemical change, the identity of the original substance changes and a new substance is formed. You must think of the physical and chemical properties of salt and of sand and come up with a procedure that allows you to separate these two materials.

You will also be required to determine the mass percent of salt in your mixture as well as the mass percent of sand. The mass percent of water in popcorn was determined in an earlier experiment. (In performing that experiment, you also separated components of a mixture. Can you identify how so?) For this salt–sand experiment, the formulas are

$$\text{mass percent of salt} = \frac{\text{mass of salt}}{\text{mass of entire sample}} \times 100$$

$$\text{mass percent of sand} = \frac{\text{mass of sand}}{\text{mass of entire sample}} \times 100$$

Because the mixture contains only two components, the total of the two percentages should be 100%.

In developing your procedure, keep in mind that you have to end up with the salt and the sand in separate containers so that you can measure the mass of each component.

Procedure

1. Develop a procedure for separating and recovering the salt and sand from a mixture of the two.
2. Have your procedure approved by your instructor (safety concerns and necessary equipment should be checked).
3. Write your procedure in the space provided on the report sheet.
4. In the space provided on the report sheet, construct a table in which you can record important data.
5. Obtain a sand–salt mixture from your instructor, record its number on the report sheet, and then carry out your procedure on this mixture, recording all data in the table you designed. (**Note:** Consider using only half of your sample to run the procedure, just in case you have to repeat the separation for any reason.)

Name ________________________________

Salt and Sand Report Sheet

Show calculations for all items marked with an asterisk (*).

1. Mixture number:______

2. Write out your approved procedure:

3. Construct your data table:

4. Sand mass percent* __________%

5. Salt mass percent* __________%

6. Obtain the true values for the mass percents from your instructor:

   true mass percent sand __________% true mass percent salt __________%

7. Calculate the percent error for your calculated mass percent values:

$$\text{percent error} = \frac{\text{true mass percent} - \text{experimental mass percent}}{\text{true mass percent}} \times 100$$

   percent error for sand __________% percent error for salt __________%

Questions

1. Describe at least one improvement you could make to your procedure to increase the accuracy of your result for the mass percent of sand.

2. Would that change also improve the accuracy of your result for the mass percent of salt?

3. What do you think was your most significant source of error in determining the mass percent of salt?

6 Radioactivity

Conceptual Chemistry Laboratory

Objectives

- To explore the effects of different shielding materials on beta and gamma radiation
- To determine background radiation levels
- To determine the half-life of a radioactive isotope (barium-137)

Materials Needed

Equipment
- Geiger counter with sample holder
- beta and gamma radioactive sources
- shielding materials (plastic and lead in various thicknesses)
- cesium-137 isogenerator set, including syringe and planchet
- clock with second hand
- rubber gloves

Chemicals
- HCl/NaCl solution

Discussion

Natural radiation surrounds us every day. The amount of natural radiation a person is exposed to, known as *background radiation,* varies with location and season. There is little we can do to avoid being exposed to background radiation. There are also other sources of radiation, such as fallout from nuclear testing and medical X rays. For health reasons, it is best to minimize our exposure. One way to limit exposure is to use a shield. In this experiment, you will test shielding materials for their effectiveness against beta and gamma radiation.

In the second part of the experiment, you will determine the half-life of the radioactive isotope barium-137. *Half-life* is a measure of the rate of decay and is defined as the time required for one half of a radioactive sample to decay. Each radioactive isotope has its own characteristic half-life. For example, cesium-137 has a half-life of 30 years. This means that half of an original amount of ^{137}Cs remains after 30 years. After another 30 years, one-fourth of the original amount remains. The isotope created when the cesium decays is barium-137, which has a half-life of 2.6 minutes. With such a short half-life, any sample of ^{137}Ba decays quickly.

Half-life can be determined by graphing the amount of radiation detected versus time. This is true because the amount of radiation given off by a sample is proportional to the amount of radioactive isotope present.

A Geiger counter records not only the radiation from the isotope being studied, but also background radiation. Before plotting a graph, therefore, you must correct your data for background radiation. To do this, you will first record the background radiation alone, then subtract it from all subsequent readings. This correction is especially important because the amount of radiation given off by the sample used in this experiment is only slightly higher than background.

To obtain a half-life from the graph, chose two points on the curve such that the y coordinate of one of the points is exactly twice the y coordinate of the other point. The half-life of the material is equal to the amount of time between these two points, read from the x axis. An example is given in Figure 1.

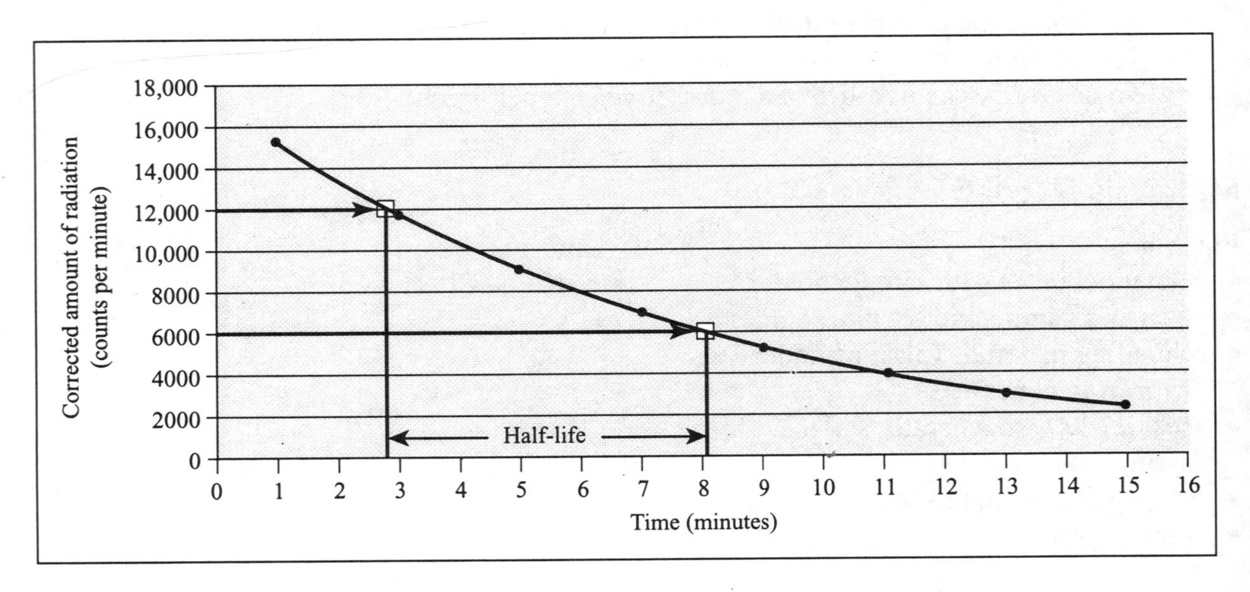

FIGURE 1 Radiation detected versus time. Readings from a Geiger counter are plotted as a function of time.

Procedure

Part A Shielding Materials

1. After obtaining various shielding materials from your instructor, fill out the first two columns of both tables in Part A of the report sheet.
2. Place a beta-emitting sample in the sample holder of the Geiger counter and hold the Geiger tube next to the sample.
3. Set the "Time" setting on the Geiger counter to 1 minute.
4. Push the "Count" button on the counter, and record the reading on the report sheet.
5. Reset the Geiger counter to zero.
6. Place various shielding materials, one at a time, directly over the sample and then hold the Geiger tube against the shielding material, as shown in Figure 2. Record the 1 minute counter reading obtained with each separate shielding material. Be sure to reset the counter with each shielding material being tested.
7. Repeat the experiment using a gamma-emitting sample.

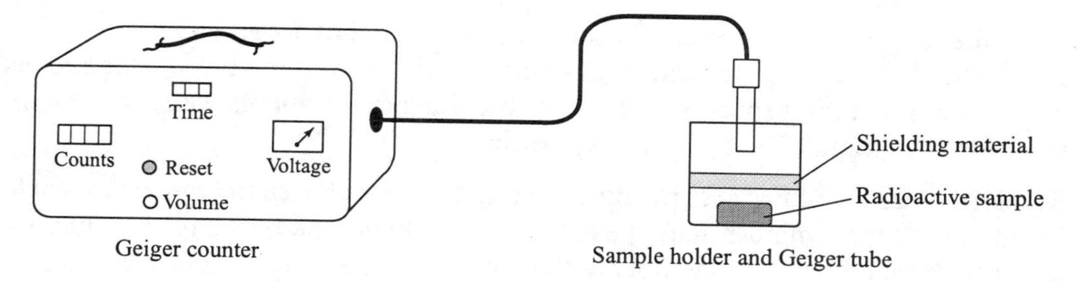

FIGURE 2

Part B Determination of Half-life

1. Set the "Time" setting on the Geiger counter to 1 minute.
2. With the sample holder empty, record the amount of radiation detected by pushing the "Counts" button and letting the Geiger counter run. The counter will automatically stop recording at the end of 1 minute. (The units will be counts per minute, cpm).
3. Push the "Reset" button on the Geiger counter, and repeat step 2 twice more. Use the average of the three trials as the background radiation, recording this average value in every line of column 3 in the table in Part B of the report sheet.
4. Fill the syringe of the isogenerator with the HCl/NaCl solution.
5. Place the planchet (small metal dish that will hold the solution containing the radioactive barium-137) under the isogenerator as shown in Figure 3.

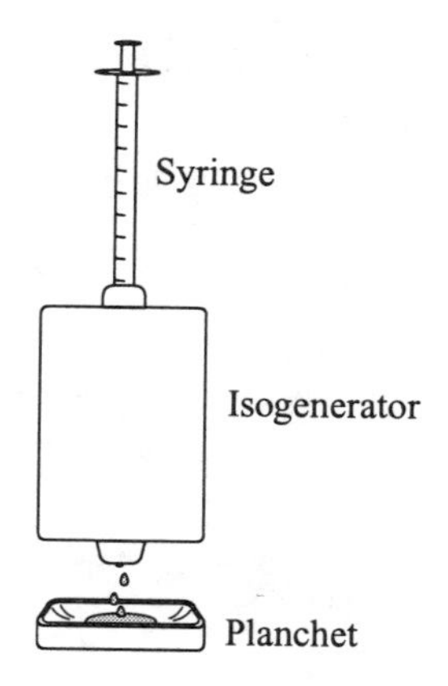

FIGURE 3

6. Attach the syringe to the top of the isogenerator, and carefully apply pressure to the syringe plunger so that the HCl/NaCl solution in the syringe is forced into the isogenerator.
7. As this solution enters the top of the isogenerator, liquid that contains your radioactive sample is forced out the bottom and into the planchet. Continue pushing down on the syringe plunger until the planchet is filled.
8. Wearing rubber gloves, carefully place the filled planchet inside the sample holder of the Geiger counter.

9. Reset the Geiger counter to zero. When the second hand of a wall clock or watch reaches 12, or after the start of a stopwatch, push the "Count" button on the counter and again let the counter run until it shuts off after 1 minute. Record the counts value in column 2 of the "Time = 1 minute" line on your report sheet.
10. Reset the Geiger counter to zero while waiting for your chosen time piece to reach 3 minutes. At the 3 minute mark, push the "count" button on the Geiger counter to let it run for another 1 minute. Record the counts value in column 2 of the "Time = 3 minutes" line of your report sheet.
11. Continue in this way—resetting the counter to zero, and starting it to run for 1minute—either until you have taken readings at 5, 7, 9, and 11 minutes or until the amount of radiation detected is near the background level, whichever comes first.
12. Correct all radiation values in column 2 of the table for background radiation, and enter the corrected values in column 4.
13. Transfer the column 4 data to the graph given on the report sheet, and connect the data points with a smooth curve.
14. Determine the half-life of your sample from your graph, using the procedure outlined in Figure 1.
15. Follow your instructor's directions for emptying the planchet (wear rubber gloves when doing so).

Name ______________________________

Radioactivity Report Sheet

Part A Shielding Materials

1. Shielding against beta emitters

Shielding Material	Thickness (cm)	Counter Reading
None	—	

2. Shielding against gamma emitters

Shielding Material	Thickness (cm)	Counter Reading
None	—	

Part B Determination of Half-life

Show calculations for all items marked with an asterisk (*).

1. Background readings

Trial 1 __________ cpm   Trial 2 __________ cpm   Trial 3 __________ cpm

Average __________ cpm

2. Radiation detected

Time (min)	Amount of Radiation from Sample (cpm)	Average Background Radiation (cpm)	Corrected Amount of Radiation from Sample (cpm)
1			
3			
5			
7			
9			
11			

*Corrected amount of radiation = amount of radiation – average background radiation.

Name ______________________________

3. Graph corrected amount of radiation versus time.

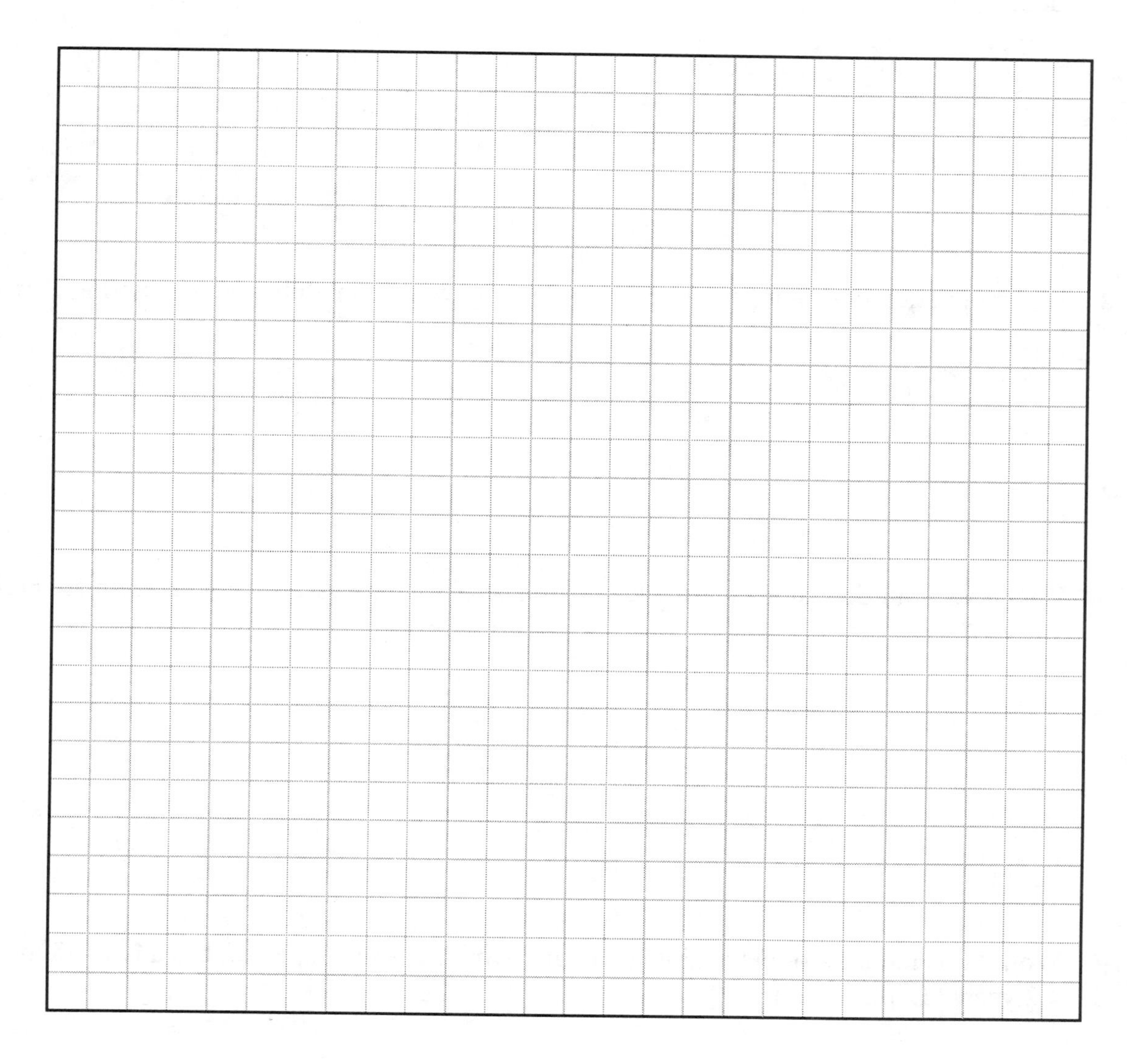

4. Half-life of sample* ________________ minutes

5. Obtain the true value of the half-life from your instructor, and calculate the percent error for your results.*

Questions

1. In Part A, which shielding material was most effective for beta radiation? For gamma radiation?

2. Alpha radiation does not penetrate glass. So, why wasn't an alpha emitter used in Part A?

3. Why does your dentist lay a large lead apron across your torso when your teeth are being x-rayed?

4. Would this method for determining half-life work for an isotope that has a half-life of 4.2 years? Explain your answer.

7 Bright Lights

Objectives

- To examine the light emitted by various elements when heated
- To identify the positive ion in an unknown metal salt by examining the light emitted by the heated salt

Materials Needed

Equipment

- Bunsen burner
- heatproof glove or potholder
- metal spatulas or flame test wires
- diffraction gratings
- spectroscope (commercial or homemade)
- ring stand with clamp large enough to fit around spectroscope barrel
- colored pencils
- discharge tubes containing various gases

Chemicals

- 0.1 *M* HCl solution
- lithium chloride powder
- sodium chloride powder
- potassium chloride powder
- calcium chloride powder
- strontium chloride powder
- barium chloride powder
- cupric chloride powder

Discussion

When materials are heated, their elements emit light. The color of the light is characteristic of the type of elements in the heated material. Lithium, for example, emits red light and copper emits green light. An element emits light when its electrons make a transition from a higher energy level to a lower energy level. Every element has its own characteristic pattern of energy levels and therefore emits its own characteristic pattern of light frequencies (colors) when heated.

It is interesting to look at the light emitted by elements through either a diffraction grating or a prism. Rather than producing a continuous spectrum of colors, the elements produce a spectrum that is discontinuous, showing only particular colors (frequencies). When the emitted light passes first through a thin slit, and then through the grating or prism, the different colors appear as a series of vertical lines, as shown in Figure 1. Each vertical line corresponds to a particular energy transition for an electron in an atom of the heated element. The pattern of lines, referred to as an *emission spectrum,* is characteristic of the element and is often used as an identifying feature—much like a fingerprint. Astronomers, for instance, can tell the elemental composition of stars by examining their emission spectra.

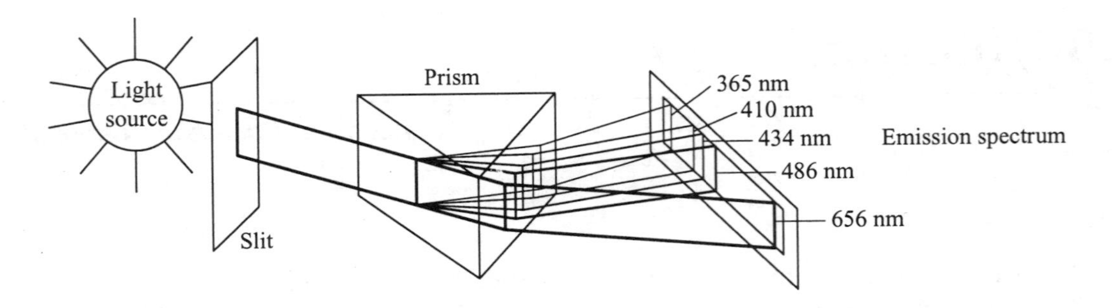

FIGURE 1 When heated, an ionized or gaseous element produces a discontinuous emission spectrum.

Procedure

Part A Flame Tests

1. Using either a heatproof glove or a potholder, hold the tip of a metal spatula in a Bunsen burner flame until the spatula tip is red hot and then dip it in a 0.1 *M* HCl solution. Repeat this cleaning process several times until you no longer see color coming from the metal when it is heated.
2. Obtain small amounts of the metal salts to be tested, and label each sample. Dip the spatula tip first into the HCl solution and then into one of the salts, so that the tip becomes coated with the powder. Then put the tip into the flame and observe the color. Record your observations on the report sheet.
3. Rinse the spatula in water, and then clean as described in step 1.
4. Repeat this procedure for all the salts, being sure to clean the spatula each time.

Part B Flame Tests Using a Spectroscope

1. Use the same procedure as in Part A, this time observing the flame through a spectroscope mounted on a ring stand. You can use either a commercial unit or a homemade one like the one shown in Figure 2.
2. On your report sheet, sketch the predominant lines you observe for each salt, using colored pencils. You will see lines both to the left and to the right of the slit. Sketch only the lines to the right. (**Note:** some salts will also show regions of continuous color.)
3. Obtain an unknown metal salt from your instructor and record its number.
4. Observe and sketch the line spectrum for your unknown, using colored pencils.
5. Identify your unknown based on its spectrum.

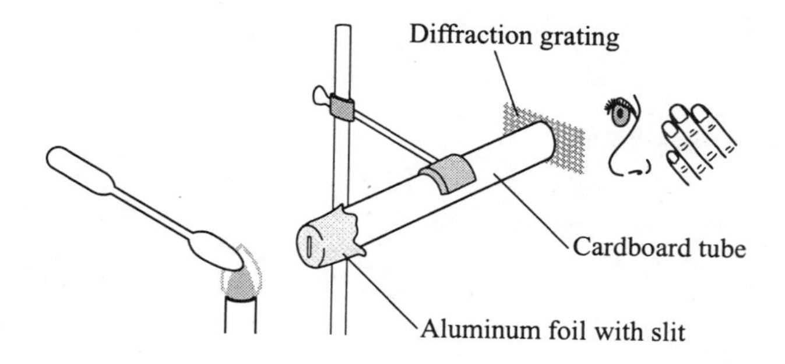

FIGURE 2 A homemade spectroscope.

Part C Gas Discharge Tubes

1. Observe through a diffraction grating the light emitted from various gas discharge tubes. (You need not pass the light through a slit because the discharge tubes themselves are narrow.)
2. Using colored pencils, sketch the line spectra for all samples available, especially hydrogen, oxygen, and water vapor.

Name ______________________________

Bright Lights Report Sheet

Part A Flame Test

Compound	LiCl	NaCl	KCl	$CaCl_2$	$SrCl_2$	$BaCl_2$	$CuCl_2$
Color							

Part B Flame Tests Using a Spectroscope

Unknown number ____________________

	Sketch of line spectrum
LiCl	
	V I B G Y O R

	Sketch of line spectrum
NaCl	
	V I B G Y O R

	Sketch of line spectrum
KCl	
	V I B G Y O R

	Sketch of line spectrum
$CaCl_2$	
	V I B G Y O R

	Sketch of line spectrum
$SrCl_2$	
	V I B G Y O R

	Sketch of line spectrum
$BaCl_2$	
	V I B G Y O R

	Sketch of line spectrum
$CuCl_2$	
	V I B G Y O R

	Sketch of line spectrum
Unknown Number	
	V I B G Y O R

Identity of unknown ________________

Part C Discharge Tubes

Substance	Sketch of line spectrum
	V I B G Y O R

Substance	Sketch of line spectrum
	V I B G Y O R

Substance	Sketch of line spectrum
	V I B G Y O R

Substance	Sketch of line spectrum
	V I B G Y O R

Substance	Sketch of line spectrum
	V I B G Y O R

Substance	Sketch of line spectrum
	V I B G Y O R

Substance	Sketch of line spectrum
	V I B G Y O R

Substance	Sketch of line spectrum
	V I B G Y O R

Questions

1. When the spatula was initially being cleaned in the flame, it may have given off yellow light. If this happened, what residue was probably on the spatula before it was cleaned?

2. What produces the colors of fireworks?

3. Is the gas in a blue "neon lamp" actually neon? Explain.

4. Does the line spectrum of water vapor bear any resemblance to the line spectra of hydrogen and oxygen? Why or why not?

8 Electron-Dot Structures

Objective

- To practice writing plausible electron-dot structures for simple molecules

Materials Needed

- periodic table

Discussion

To help predict how atoms bond together, you can use the *octet rule,* which states that atoms form chemical bonds so as to have a filled outermost occupied shell. Looking at a periodic table, we see that the noble gases already have filled outermost occupied shells, which explains why they tend not to form chemical bonds. Note that all the noble gases except helium have *eight* outermost electrons; hence the name octet rule.

The octet rule can be used to build an ammonia molecule, NH_3. Knowing that nitrogen has five valence electrons and hydrogen one, you can satisfy the octet rule with the electron-dot structure

```
    ••
H : N : H      (eight electrons surrounding N,
    ••          two electrons surrounding H)
    H
```

In this structure, the nitrogen has eight valence electrons so that its outermost occupied shell, which has a capacity for eight electrons, is filled. Each hydrogen has two valence electrons so that its outermost occupied shell, which has a capacity for two electrons, is also filled.

Using the octet rule, you will learn to draw electron-dot structures of simple covalent compounds given their chemical formulas. (This activity focuses on bonding in simple covalent molecules that obey the octet rule. There are many molecules that do not follow this rule, but they are not introduced in this laboratory.)

To construct a plausible electron-dot structure for a molecule,

1. Determine the total number of valence electrons available from the chemical formula by adding up the valence electrons of all atoms in the molecule.
2. Write the chemical symbols for all atoms in the molecule, arranging the symbols as you think they might appear in the molecule. Many molecules contain a central atom. If there are many elements present, place them in the order in which they are written in the formula.
3. Place single bonds between all pairs of atoms, remembering that each bond represents *two* electrons, one from each atom.
4. Add the appropriate number of electrons around each atom so that the atom obeys the octet rule (eight electrons around all atoms other than hydrogen, two around hydrogen).

5. Count the number of electrons used in your structure. If the number is equal to the sum you calculated in step 1, your structure is plausible and you are done. If the number of electrons used is greater than the sum you calculated in step 1, try putting in multiple bonds (try double first, then triple) until all atoms obey the octet rule and the number of electrons used matches the number calculated in step 1.
6. If the number of electrons used is less than the sum you calculated in step 2, you probably made an error in your count in step 1 and so should recalculate.

Examples

To show these steps in action, here are examples using carbon tetrabromide and carbon monoxide.

1. Total number of valence electrons available:

Carbon tetrabromide (CBr_4)		Carbon monoxide (CO)	
1 C atom (4 valence electrons each)	1 × 4 = 4	1 C atom (4 valence electrons)	1 × 4 = 4
4 Br atoms (7 valence electrons each)	4 × 7 = 28	1 O atom (6 valence electrons)	1 × 6 = 6
	Total	32	
Total	10		

2. Arrangement of symbols for all atoms (locate central atom if applicable):

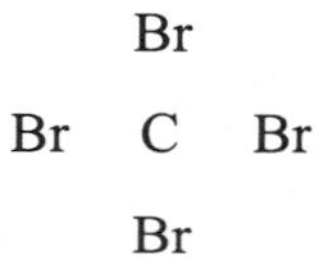

```
     Br
Br   C   Br                    C   O
     Br
```

3. Single bonds between all pairs of atoms:

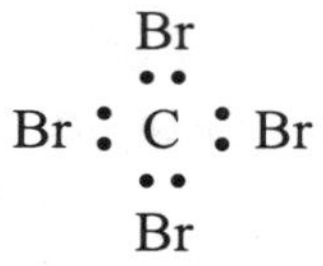

```
      Br
      ••
Br  : C :  Br                  C : O
      ••
      Br
```

4. Add remaining electrons to complete octets:

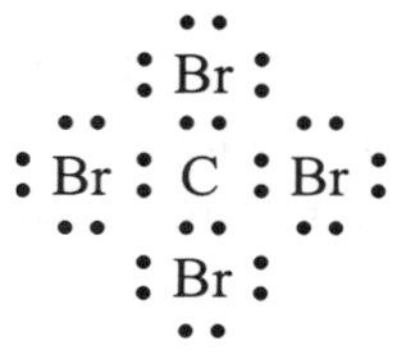

```
        ••
       :Br:
   ••   ••   ••                ••  ••
  :Br : C : Br:               :C : O:
   ••   ••   ••                ••  ••
       :Br:
        ••
```

5. Count the number of electrons used:

The above structure uses 32 electrons, which is equal to the number calculated in step 1. The structure is complete.

The above structure uses 14 electrons, more than the total number calculated in step 1. A multiple bond must be present. With a double bond, the structure would be

:C::O:

Still there is a problem because the structure uses 12 electrons. A triple bond must be present:

:C⋮⋮O:

Finally, the structure obeys the octet rule and uses the total number of valence electrons available. The structure is complete.

Many chemists prefer to use a line to represent a bonded pair of electrons rather than dots. With this line notation, these two molecules are represented as

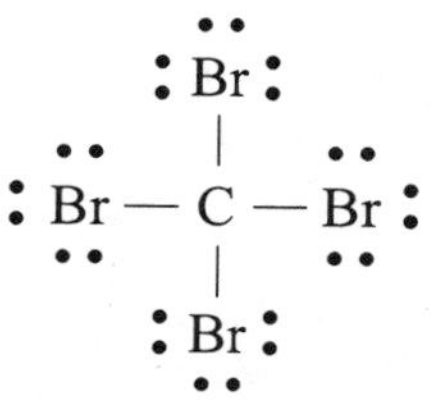

```
       ..
      :Br:
       |
 ..         ..
:Br — C — Br:
 ..         ..
       |
      :Br:
       ..
```

:C≡O:

CBr4 e⁻ dot structure:
1) count # of valence e
2) Identify the central atom – single, double and triple bonds

9 Molecular Shapes

Objectives

- To become familiar with the three-dimensional shapes of molecules
- To build molecular models from information given in electron-dot structures
- To draw electron-dot structures from information given in molecular models
- To predict the polarity of a molecule from its molecular shape

Materials Needed

- molecular model kit
- built models of six different molecules

Discussion

It is extremely useful for a chemist to know the three-dimensional shape of the molecules of a given compound. For some compounds (those containing two or more polar bonds), a knowledge of molecular shape is necessary when predicting polarity. One can illustrate the relationship between molecular shape and polarity using water as an example. The O–H bond is polar because of the difference in the electronegativities of the two elements. A water molecule contains two O–H bonds. If a water molecule were linear, as shown in Figure 1, the two polar bonds would be equal in magnitude but opposite in direction. Their effects would cancel, and as a result the water molecule as a whole would be nonpolar. Water molecules have, however, a bent shape as shown in Figure 2. In such a structure, the two polar bonds do not cancel because they do not point in opposite directions. As a result, the water molecule is polar.

FIGURE 1 FIGURE 2

We shall use valence shell electron-pair repulsion theory (VSEPR) as our guide in predicting molecular shapes. The basis of this theory is that pairs of electrons surrounding an atom try to get as far away from one another as possible to lessen electrostatic repulsion. In order to apply this theory, we must know how many pairs of electrons are around each atom in a molecule. It is therefore necessary to begin with the electron-dot structure for the molecule.

Table 1 shows possible molecular shapes derived from VSEPR theory. Note how the electron pairs are placed as far apart as possible, whether the pair is a bonding pair or a nonbonding pair. Molecular *geometry* depends on the total number of substituents about the central atom. A *substituent* is defined as any atom or any nonbonding pair of electrons. Notice that although the total number of electrons around the central atom is always eight (four pairs), the number of substituents can be less than four.

Remember that according to the VSEPR model, pairs of valence electrons strive to get as far apart as possible. With a multiple bond, the pairs cannot separate from one another because they are all being shared by the same two nuclei. For the purposes of VSEPR, therefore, a multiple bond is treated as a single bond and the noncentral atom taking part in the bond counts as *one* substituent.

Molecular *shape* depends on the placement of *atom substituents only;* nonbonding pairs are *not* considered when you are determining molecular shape. When there are no nonbonding pairs of electrons about a central atom, the molecular shape is identical to the molecular geometry. It is only when nonbonding pairs are present that a molecule's shape differs from its geometry.

Procedure

Part A From Electron-Dot Structure to Model

Using the molecular model kit provided by your instructor, build a model of each molecule shown in Table 2 of the report sheet. Then sketch your molecules in the appropriate spaces in Table 2, using the bond notation (—, ·······, ▬) used in column 5 of Table 1. Then complete columns 3 and 4 of Table 2.

Part B From Model to Electron-Dot Structure

Use the three-dimensional models provided by your instructor to complete Table 3 of the report sheet.

TABLE 1

General Lewis Structure	Number of Substituents	Number of Nonbonding Pairs	Molecular Geometry	Molecular Shape
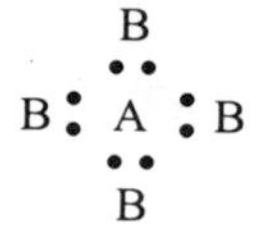 B : A : B (B above and below; four B)	4	0	 109° Tetrahedral	 Tetrahedral
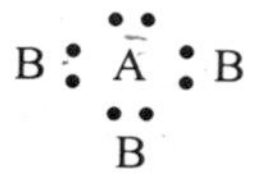 B : A : B (B below; one nonbonding pair)	4	1	 109° Tetrahedral	 Triangular pyramidal
 B : A : B (two nonbonding pairs)	4	2	 109° Tetrahedral	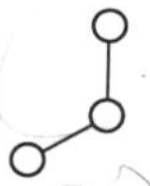 Bent
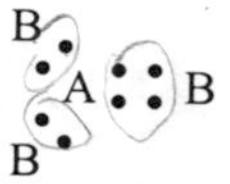 B, A : B, B (three B)	3	0	 120° Triangular planar	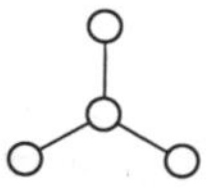 Triangular planar
 A : B, B (one nonbonding electron pair)	3	1	 120° Triangular planar	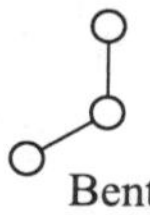 Bent
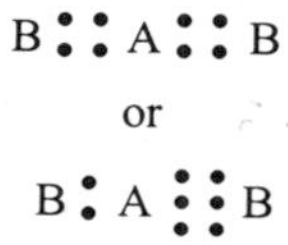 B :: A :: B or B : A ::: B	2	0	 180° Linear	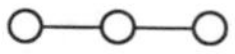 Linear

10 Solutions

Objectives

- To create and seed a supersaturated solution
- To explore the effect of temperature, particle size, and stirring on the rate of dissolving
- To make and calibrate a hydrometer
- To determine the sugar content of various beverages

Materials Needed

Equipment
- hot plate
- ring stand and clamp
- 400-mL beaker
- five medium test tubes
- glass stirring rod
- wooden test tube rack
- mortar and pestle
- 9-inch plastic pipet
- small hardware nuts for weighting down pipet
- 50-mL graduated cylinder
- metric ruler calibrated in millimeters

Chemicals
- sodium acetate trihydrate crystals
- rock salt crystals
- 4%, 8%, 12%, 16% sugar solutions
- various soft drinks and fruit juices

Discussion

A solution is any homogeneous mixture. The *solvent* in a solution is the component present in the greatest amount, and the component(s) present in lesser amounts are referred to as *solute(s)*. The *solubility* of a solute in a given solvent is a measure of how well the solute dissolves in the solvent. For most solutes, solubility is a function of temperature. A solution is saturated when it contains the maximum amount of solute, unsaturated when it contains less than the maximum, and supersaturated when it contains more than the maximum. In this experiment, you will look at how temperature affects solubility by preparing a supersaturated solution.

The rate at which a solute dissolves depends on the amount of contact between solute and solvent particles. You will explore how temperature, stirring, and particle size affect solution rate by dissolving rock salt in water under a variety of conditions.

The concentration of a solution is a measure of the amount of solute in a given volume of the solution. There are many units used to express concentration, including molarity (moles of solute per liter of solution) and percent by mass (mass of solute per 100 milliliters of solution). Although molarity is a favorite among chemists, you will work with percent by mass in this experiment.

Most soft drinks and fruit juices are sugar solutions. The concentration of sugar in the beverages can be found by using a *hydrometer* (Figure 1), which is a flotation device that measures the density of a liquid. The greater the density of the liquid, the higher the hydrometer floats. For sugar solutions, the greater the sugar concentration, the greater the density and hence the higher the hydrometer floats. You will first prepare a calibration curve showing the hydrometer height versus the concentration of various sugar solutions. You will then use your calibration curve to determine the sugar content of some soft drinks and fruit juices.

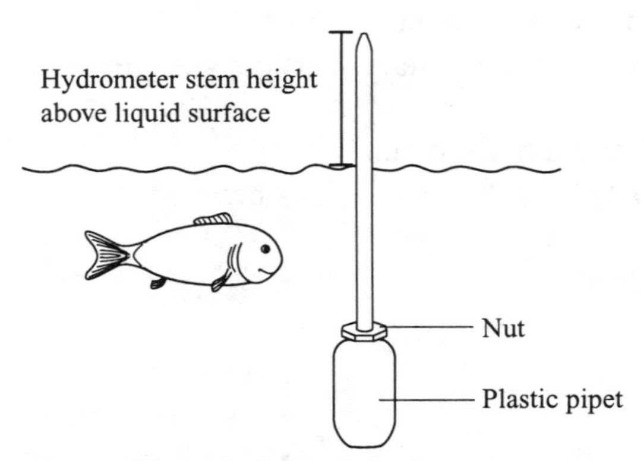

FIGURE 1 A simple hydrometer.

Procedure

Part A Creating and Seeding a Supersaturated Solution

1. Place the hot plate on the base of the ring stand as shown in Figure 2.
2. Fill the 400-mL beaker about two-thirds full with water and place it on the hot plate set to medium.
3. Bring the water to a boil. (You may need to raise the heat setting to get the water to boil.)
4. Add sodium acetate trihydrate crystals to one of the test tubes until the tube is about one-quarter full.
5. Add enough water to just cover the crystals.
6. Using the clamp of the ring stand, suspend the test tube in the boiling water bath as shown in Figure 2.
7. Use the glass stirring rod to stir the contents of the test tube until all the crystals are dissolved.
8. Raise the clamp so that the test tube is out of the water bath.
9. Turn off the hot plate.
10. Let the solution in the test tube cool undisturbed to room temperature.
11. Add a small crystal of sodium acetate trihydrate to the test tube, and record what happens.

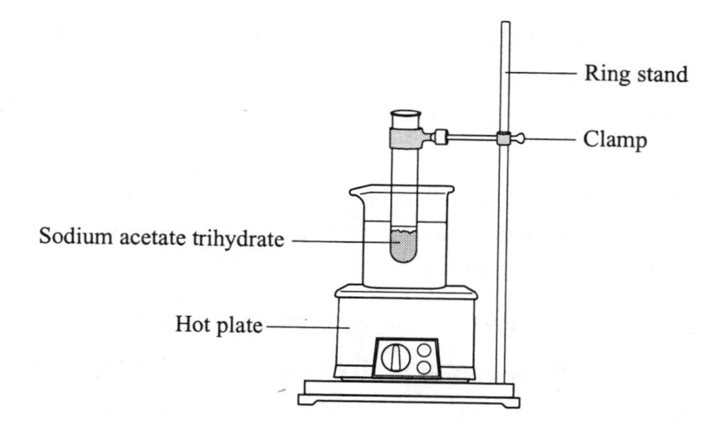

FIGURE 2 Setup for dissolving sodium acetate trihydrate.

Part B Effect of Temperature, Stirring, and Particle Size on Solution Rate

1. Number four test tubes 1, 2, 3, and 4 and place them in the test tube rack. Fill each about one-half full with water.
2. Obtain four pea-sized crystals of rock salt, and grind one of them to a fine powder in the mortar.
3. To tube 1, add a single crystal of rock salt and record the amount of time it takes for the crystal to dissolve. Do not shake or stir the solution.
4. Reheat the water bath used in Part A to boiling, adding water to the beaker if necessary.
5. Place tubes 2, 3, and 4 in the boiling water bath, and let the water in the tubes heat for about 5 minutes.
6. To tube 2, add a single crystal of rock salt and record the amount of time it takes for the crystal to dissolve. Do not shake or stir the solution.
7. To tube 3, add a single crystal of rock salt and stir until the crystal dissolves. Record the amount of time it takes for the crystal to dissolve.
8. To tube 4, add the remaining ground rock salt and stir until it completely dissolves. Record the amount of time it takes for the powder to dissolve.

Part C Construction and Calibration of Simple Hydrometer

1. Fill the 9-inch pipet approximately one-half full with water and invert. All the water should run into the bulb.
2. Slip a nut onto the stem end and allow it to rest on the "shoulders" of the bulb, as shown in Figure 1.
3. Test your hydrometer by placing it, bulb end down, in the 50-mL graduated cylinder containing 50 mL of water. The pipet should float with about 2.5 cm of the stem sticking out above the water. If it sticks out much more or less than this, either add water to the pipet or remove water from it.

4. When you have adjusted the amount of water in the pipet bulb so that the stem sticks out about 2.5 cm, measure the height of the stem above the surface of the water in the graduated cylinder. Measure to the nearest millimeter (0.1 cm), and record the height on your report sheet.
5. Remove the hydrometer from the graduated cylinder. *Being careful not to let any water spill out of the hydrometer or any water get in,* rinse the outside surface of the hydrometer well and dry completely.
6. Empty out the graduated cylinder, rinse well, and dry completely.
7. Add 50 mL of the 4% sugar solution to the graduated cylinder, place the hydrometer in the solution—remember, bulb end down—and measure and record the height of the stem above the surface of the solution.
8. Repeat steps 5, 6, and 7 for 8%, 12%, and 16% sugar solutions.
9. On the blank graph in Part C of your report sheet, label the horizontal axis with numbers representing the percentages of the sugar solutions. (Two squares equalling 1% would be a reasonable scale.) Label the vertical axis with numbers that represent various stem heights for the hydrometer. Choose your scale for this axis so that the lowest number you write is a bit lower than the lowest height you measured and the highest number you write is a bit higher than the highest height you measured.
10. Plot your measured heights on this graph. This is your calibration curve, and a sample one is shown in Figure 3.

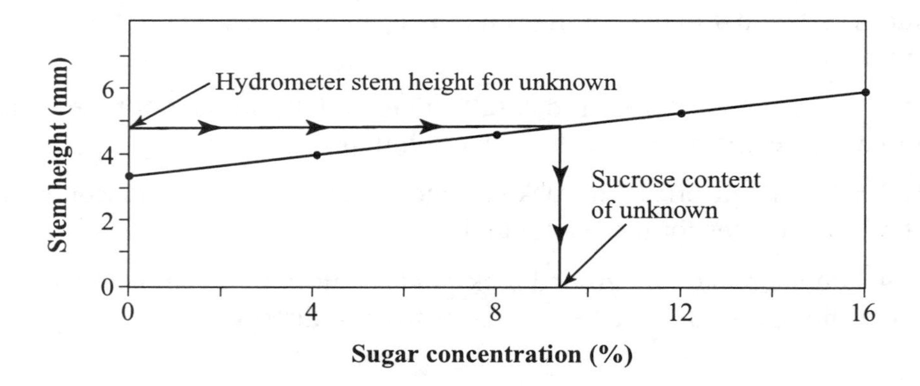

FIGURE 3 Sample of calibration curve for determining sugar concentration. Find the point on the vertical axis that corresponds to the hydrometer stem height for your unknown solution (4.8 mm in this graph). Draw a horizontal line that runs rightward from this point until it intersects the calibration curve. This line is shown here as →→. Now drop a vertical line from the intersection point down to the horizontal axis, shown here as ↓↓. The value at the point where this line intersects the horizontal axis—about 9.4% here—is the sugar concentration of your unknown solution.

Part D Determination of Sugar Content in Beverages

1. Obtain soft drink and/or juice samples from your instructor. If you are given a carbonated drink (your instructor will tell you), you must remove the carbonation before running this experiment. To do this, empty out the beaker used for the water bath in part A, pour about 60 mL of one of your carbonated samples into the beaker and heat to boiling on the hot plate. Let the solution boil for 3 or 4 minutes, and then cool to room temperature. (This step must be done because with a carbonated solution, bubbles will collect on your hydrometer bulb and affect its buoyancy.
2. Following the procedure of Part C, measure and record the stem height for each of your unknown samples.
3. Use the calibration curve you prepared in Part C for your particular hydrometer to determine the sugar concentration in each of your unknowns.

11 Candy Chromatography

Objectives

- To isolate dyes from candies
- To separate the components of dyes in various candies
- To compare the composition of dyes used in different brands of the same color of candy

Materials Needed

Equipment

- 400-mL beaker
- hot plate
- three medium test tubes
- 15-cm piece of woolen yarn (pale in color)
- stirring rod
- red litmus paper
- evaporating dish
- scrap strip of chromatography paper
- capillary tubes
- sheet of chromatography paper, 14 cm × 10 cm
- pencil
- ruler
- stapler
- 1-L beaker
- plastic food wrap or aluminum foil

Chemicals

- candy (M&Ms, Skittles)
- vinegar
- 1.0 *M* ammonia solution
- standard solutions for all colors represented in candies used
- FD&C food dyes

Discussion

Many items, from clothing to food, are dyed with chemical dyes that contain a combination of pigments. In this experiment, you will use paper chromatography to separate into components the dyes used in M&M and Skittles candies. You will then compare these components against food color standards.

Chromatography is a separation technique that takes advantage of differences in the affinities the various components of a mixture have for different mediums. For all types of chromatography, there are at least two mediums in different phases—one that is called the stationary phase, often a solid or a liquid, and another that is called the mobile phase, often a liquid or a gas.

In paper chromatography, paper is the stationary phase. The mixture to be separated is spotted onto the paper, and the paper is placed upright in a liquid solvent, as shown in Figure 1. The solvent is allowed to travel up the paper, meaning that in this case the mobile phase is a liquid. The components of the mixture move various distances up the paper. In general, the greater the affinity a component has for the liquid, the faster that component travels along with the liquid. Conversely, the greater the affinity a component has for the stationary paper, the slower it travels along with the liquid. The components of a mixture can therefore be separated from one another as they travel along with the solvent at different rates. This process results in a series of dots known as a chromatogram. A sample is shown in Figure 2.

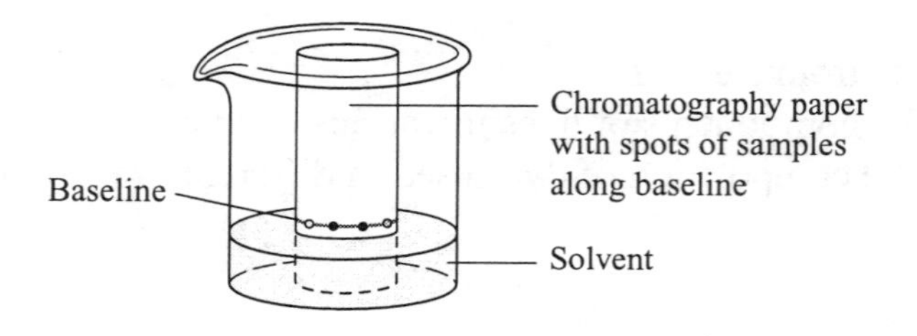

FIGURE 1

Here is a picture of a chromatogram before it is developed:

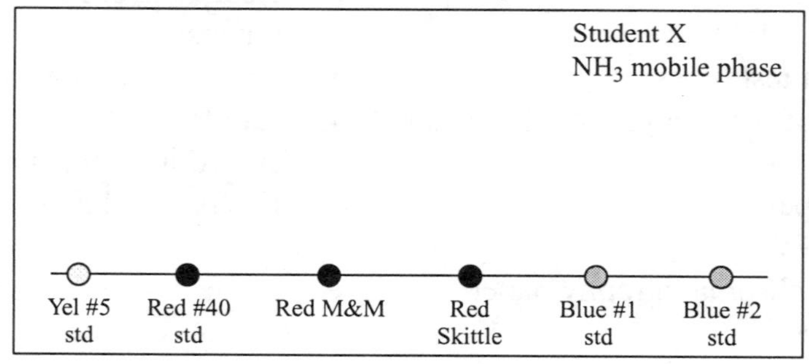

Here is a hypothetical picture of the same chromatogram **after** it is developed:

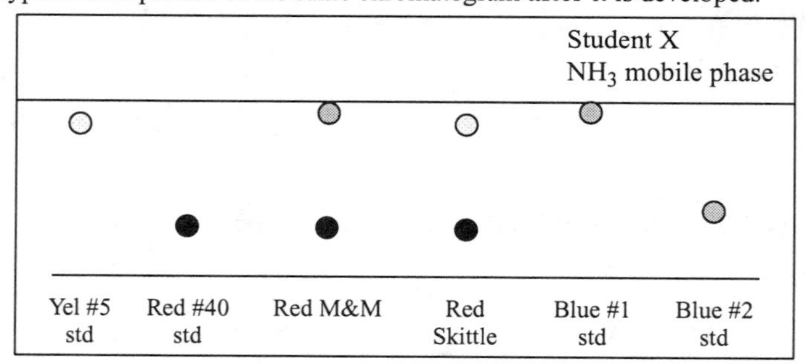

FIGURE 2

Procedure

You'll be working with a partner who will extract the dye from candy that is the same color as yours but a different brand. For instance, if you are doing blue M&Ms, you should be working with a person doing blue Skittles. Each of you will extract the dye from the one brand of candy, but then you'll each run your own chromatogram on both blue dyes.

Part A Extracting the Dye

1. Add about 150 mL of water to a 400-mL beaker.
2. Heat the water to boiling on a hot plate set to medium (adjust the heat to keep the water just at the boiling point). Add more water if the level gets too low as the experiment progresses.
3. Place six candies (all of the same color and same brand) in a test tube, and then add enough vinegar to cover them.
4. Heat the test tube in your boiling water bath until the colored coating of the candy has dissolved (avoid dissolving the interior of the candy).
5. Remove the test tube and allow it to cool to room temperature. Your solution now contains dyes, some sugar, and vinegar.
6. Pour the liquid from your test tube into a clean test tube, leaving behind all solids.
7. Add the 15-cm length of woolen yarn and 3 mL of vinegar to the test tube containing the dye solution.
8. Heat this tube in the boiling water bath for about 5 minutes, stirring occasionally.
9. Remove the yarn and rinse it with a little tap water (don't rub it). The dye should now be stuck to the yarn.
10. Place the yarn and about 5 mL of the 1.0 *M* ammonia solution in a clean test tube, and mix with a stirring rod.
11. Test the solution to make sure it is basic by placing a drop of it on a piece of red litmus paper. If the paper stays red, add a few more drops of ammonia to the test tube and retest. Once the paper turns blue, continue to the next step.
12. Heat the test tube containing the yarn and ammonia in the boiling water bath for about 5 minutes, stirring occasionally.
13. Remove the yarn and pour the solution into an evaporating dish.
14. Heat the evaporating dish gently on the hot plate set to low to concentrate the solution. Stop just short of dryness—DO NOT let all the liquid evaporate.

Part B Preparing the Chromatogram

1. Using the scrap piece of chromatography paper and a capillary tube, practice making small spots of dye solution on the paper.
2. On the 14 × 10 chromatography paper, draw a straight baseline *in pencil*—do not write on the paper with ink—approximately 1.5 cm up from the bottom of the sheet. (Consider the 14-cm edges to be the top and bottom of the sheet.)

3. Using separate capillary tubes for each solution, place one spot of each of the following solutions along your baseline:

   the dye solution you prepared

   the dye solution of the same color but from a different brand of candy, prepared by a partner

   separate spots of the standard solutions for the color you are working with (your instructor will tell you which standard solutions to use)

   In applying the spots, you should follow these guidelines:

   Use a fresh capillary tube for each solution.

   Place the first spot at least 3 cm in from the left edge of the paper and the last spot at least 3 cm in from the right edge of the paper.

   Leave about 2 cm of space between neighboring spots.

4. In *pencil* (*not* ink), label each spot along the bottom edge of the paper.
5. Holding the paper upright with the spots facing you, curve the two vertical edges away from you and toward each other until they just touch each other (make sure they do not overlap). Staple the touching sides together at the top and bottom of the cylinder you've created.

Part C Developing the Chromatogram

1. Add about 40 mL of the 1.0 *M* ammonia solution to the 1-L beaker (not more than 0.5 cm deep). Then stand your paper cylinder alongside the beaker and check to see that the level of the ammonia in the beaker is below the baseline on the cylinder. If the ammonia level is above the baseline, pour out some of the solution and check again.
2. Once you are sure your spots won't be submerged, place the cylinder in the beaker, baseline at the bottom, as shown in Figure 1.
3. Cover the beaker with plastic wrap or aluminum foil. Leave the setup undisturbed once the solvent has begun moving up the paper.
4. When the solvent has traveled to just below the top of the paper cylinder, remove the cylinder from the beaker and allow it to dry.

Name ______________________________

Candy Chromatography Report Sheet

Data

Attach your chromatogram (or draw a diagram of it) here.

Questions

1. Were the same pigments present in the two different brands of candy?

2. Chromatography paper is fairly polar. How would you expect an ionic component to behave when the mobile phase is a relatively nonpolar solvent, such as acetone? Would such a component readily travel with the solvent or would it stay behind?

3. Which type of intermolecular forces would be responsible for the movement of an ionic component?

4. How would you expect a relatively nonpolar component to behave when the mobile phase is a nonpolar solvent? Explain your reasoning.

12 How Much Fat?

Objectives

- To extract fat from potato chips
- To compare the mass percentage of fats in regular and low-fat potato chips

Materials Needed

Equipment

- two 25-mL Erlenmeyer flasks
- balance
- mortar and pestle
- two 13 × 100 mm test tubes with stoppers
- 5-mL graduated cylinder
- 5-mL pipet
- 100-mL beaker
- hot plate

Chemicals

- regular potato chips
- low-fat potato chips
- petroleum ether

Discussion

We have all wondered if it is worth eating low-fat or even nonfat chips. In this experiment, you will determine the fat content of various snack chips. The process you will use to get the fat from the chips, called extraction, is a separation technique based on solubility. The fats are separated from other components based on polarity. Fats are nonpolar and therefore soluble in nonpolar solvents, such as petroleum ether. Assuming that all other components in the chips are polar, only the fats should dissolve into a nonpolar solvent.

You will determine the mass of the fat you extract and then compare the mass percentage of fats in regular and low-fat chips. You will then calculate the efficiency of the extractions by comparing your experimental values with the values stated on the packaging of the chips.

Procedure

1. Determine the mass of an empty 25-mL Erlenmeyer flask and record the mass in the "Regular Chips" column in the report sheet.
2. Use the pestle to crush the regular potato chips in the mortar.
3. Determine and record the mass of approximately 0.5 gram of the crushed chips and transfer the sample to one of the test tubes.
4. Using the graduated cylinder, add 3 mL of petroleum ether to the test tube.
5. Stopper the test tube and shake for several seconds. Periodically remove the stopper to relieve any pressure that may build up.
6. Let the solid in the test tube settle to the bottom.

7. Remove the stopper and pipet the liquid into the pre-weighed Erlenmeyer flask. Do not allow any solid to be transferred to the flask.
8. After the original 3 mL of petroleum ether has been removed (or as much as possible removed), add another 3 mL of petroleum ether to the test tube, stopper and shake for several seconds, then allow the solid to settle to the bottom.
9. Remove the stopper and pipet the liquid into the same Erlenmeyer flask. Again, do not transfer any solid.
10. Place approximately 25 mL of water in the 100-mL beaker and place the beaker on a hot plate set to medium under a fume hood.
11. Place the Erlenmeyer flask containing the petroleum ether and fats in the hot water. Be very careful not to let any water get into the Erlenmeyer flask.
12. After all the petroleum ether has evaporated (usually about 10 to 15 minutes), allow the flask to cool to room temperature.
13. The solid residue in the flask is the fat you extracted from the chips. Weigh and record the mass of the flask plus fat.
14. Repeat the procedure using low-fat chips as your sample.

13 Energy and Calorimetry

Objectives

- To identify a metal by determining its specific heat capacity
- To compare the energy content per gram of various fuels

Materials Needed

Equipment
- hot plate
- ring stand with clamp large enough to hold a 400-mL beaker
- two 400-mL beakers
- balance
- string
- two large Styrofoam cups
- 25-mL graduated cylinder
- thermometer
- scissors
- tongs
- matches

Chemicals
- chunk of unknown metal
- candle
- can of Sterno fuel

Discussion

Energy is defined as the capacity to do work. Energy is often measured in calories (cal), and calorimetry is the science of measuring calories. One calorie equals 4.184 joules and is defined as the amount of energy required to raise the temperature of 1 gram of water by 1 Celsius degree. The amount of heat energy necessary to raise the temperature of 1 gram of any material by 1 Celsius degree is defined as the material's specific heat capacity (c). The specific heat capacity of water is 1 cal/g·°C = 4.184 J/g·°C.

Energy is involved in all physical and chemical changes. A change that absorbs energy from the surroundings is referred to as endothermic, and a change that releases energy to the surroundings is exothermic. The law of conservation of energy states that energy cannot be created or destroyed. It can be transferred from one object to another, however. When all the energy is transferred and none lost to the surroundings, the amount of energy released by one object must always equal the amount of energy absorbed by the other.

The amount of energy involved in heating or cooling a sample can be calculated with the formula

$$\text{heat energy} = mc\Delta T$$

where m is the mass of the sample, c is its specific heat capacity in cal/g·°C and ΔT is the change in temperature of the sample.

In the first part of this experiment, you will use a simple calorimeter to determine the specific heat capacity of a metal. You will heat a metal sample of known mass to the temperature of boiling water (100∞C) and then place the sample in a double Styrofoam cup containing a known mass of water. If the temperature of the water is taken before and after the metal is transferred, the following formulas can be used to determine the specific heat capacity of the metal:

$$\text{heat energy lost by metal} = \text{heat energy gained by water}$$

$$m_{\mathrm{m}} c_{\mathrm{m}} \Delta T_{\mathrm{m}} = m_{\mathrm{w}} c_{\mathrm{w}} \Delta T_{\mathrm{w}}$$

$$c_{\mathrm{m}} = \frac{m_{\mathrm{w}} c_{\mathrm{w}} \mathrm{D}T_{\mathrm{w}}}{m_{\mathrm{m}} \mathrm{D}T_{\mathrm{m}}}$$

In the second part of the experiment, you will compare the energy content of various fuels. The mass of a fuel source (a candle or Sterno fluid) will be determined before and after the fuel is burned. The fuel will be used to heat a sample of water. The amount of heat energy released from the fuel is equal to the amount of heat energy gained by the water:

$$\text{heat energy released by fuel} = \text{heat energy absorbed by water} = m_{\mathrm{w}} c_{\mathrm{w}} \Delta T_{\mathrm{w}}$$

The fuel's energy content per gram is then calculated by dividing this amount of heat energy by the mass of fuel burned.

Procedure

Part A Specific Heat Capacity of a Metal

1. Place the hot plate on the base of the ring stand.
2. Add approximately 250 mL of water to one of the 400-mL beakers, and heat the water to boiling on the hot plate.
3. Obtain an unknown metal sample and record its number on your report sheet.
4. Determine and record the mass of your sample.
5. Tie a string around the metal sample and suspend it in the beaker of boiling water as shown in Figure 1.
6. While the metal is heating, nest one Styrofoam cup inside a second one.
7. Determine and record the combined mass of the two cups.
8. Pour into the graduated cylinder the same number of milliliters of water as there are grams in your metal sample, and then pour this water into the inner cup.
9. Determine and record the mass of the cups plus water.
10. Measure and record the temperature of the water in the inner cup.

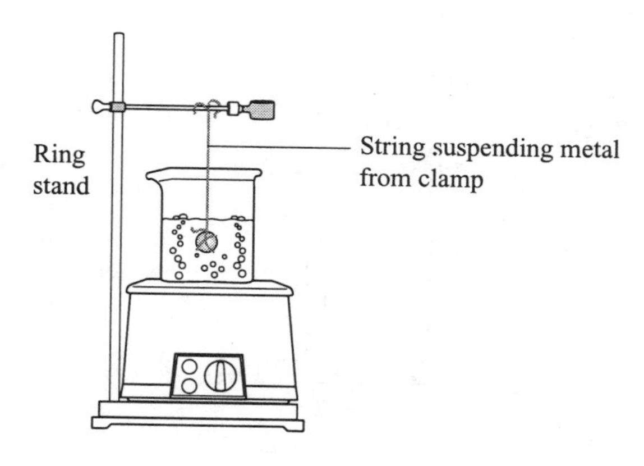

FIGURE 1

11. Cut the string holding your metal sample so that the sample falls to the bottom of the beaker. Then, using tongs, *quickly* remove the metal sample from the boiling water, shake off any excess water, and drop the heated metal into the inner Styrofoam cup.
12. Stir the water in the cup carefully with a thermometer and record the highest temperature attained by the water after the addition of the heated metal.
13. Use the formulas given in the discussion section to calculate the specific heat capacity of your sample.
14. Compare your calculated value for specific heat capacity with the values in Table 1 to determine the identity of your metal.

TABLE 1 Specific Heat Capacities

Metal	Specific Heat Capacity (cal/g·C°)
Aluminum	0.215
Copper	0.0923
Lead	0.0301
Tungsten	0.0321
Zinc	0.0925

Part B Energy Content of Fuels

1. Determine and record the mass of a dry 400-mL beaker.
2. Add approximately 100 mL of water.
3. Determine and record the mass of the beaker plus water.
4. Suspend the beaker through a utility clamp as shown by Figure 2, being sure the beaker is securely held before releasing it.
5. Determine and record the initial temperature of the water. Leave the thermometer in the water for the rest of the experiment.

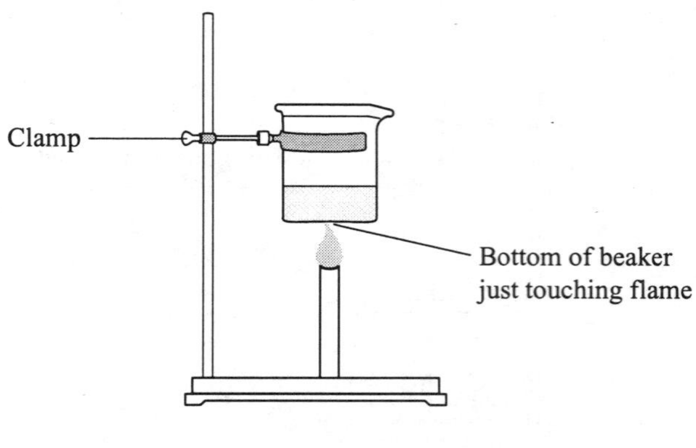

FIGURE 2

6. Determine and record the initial mass of the candle.
7. Place the candle under the suspended beaker of water, and move the clamp up or down the stand so that the bottom of the beaker just touches the wick.
8. Light the candle and let it burn until the water temperature has increased by at least 5 C° (the larger the temperature change, the better).
9. Extinguish the candle and record the final temperature of the water.
10. When the candle is cool, determine and record its final mass.
11. Repeat the experiment using a can of Sterno in place of the candle.

Name ______________________________

Energy and Calorimetry Report Sheet

Part A Specific Heat Capacity of a Metal

Show calculations for all items marked with an asterisk (*).

1. Unknown metal number ______________________________
2. Appearance of unknown metal ______________________________
3. Mass of unknown metal __________ g
4. Mass of doubled Styrofoam cups __________ g
5. Mass of doubled Styrofoam cups plus water __________ g
6. Mass of water* __________ g
7. Initial temperature of metal __________ °C
8. Final temperature of metal __________ °C
9. Temperature change of metal* __________ °C
10. Initial temperature of water in cup __________ °C
11. Highest temperature of water in cup __________ °C
12. Temperature change of water in cup* __________ °C
13. Specific heat capacity of metal* __________ cal/g·°C
14. Identity of metal ______________________________

Part B Energy Content of Fuels

Show calculations for all items marked with an asterisk (*).

	Candle	Sterno
1. Mass of empty beaker	________ g	________ g
2. Mass of beaker plus water	________ g	________ g
3. Mass of water*	________ g	________ g
4. Initial mass of fuel source	________ g	________ g
5. Final mass of fuel source	________ g	________ g
6. Mass of fuel burned*	________ g	________ g
7. Initial temperature of water	________ °C	________ °C
8. Final temperature of water	________ °C	________ °C
9. Temperature change of water*	________ °C	________ °C
10. Heat energy gained by water*	________ cal	________ cal
11. Heat energy released by fuel source	________ cal	________ cal
12. Heat energy released per gram of fuel burned*	________ cal/g	________ cal/g

Questions

1. Could two metal cups be used in place of the Styrofoam cups? Why or why not?

Name ______________________________

2. In Part A, the metal sample tends to change temperature much more rapidly than the water. Does this indicate that the specific heat capacity of the metal is higher than that of water or lower?

3. Based on your data, do all fuel sources contain the same amount of energy per gram?

4. Some of the heat generated by the burning candle in Part B is lost to the surroundings rather than being absorbed by the water. If this heat loss were taken into account, would your calculated heat energy released per gram of fuel burned be greater or less than the value you reported?

14 The Clock Reaction

Objective

- To observe how catalysts, reactant concentration, and temperature affect the rate of a chemical reaction

Materials Needed

Equipment
- two 25-mL graduated cylinders
- fourteen 50-mL beakers
- eyedroppers
- stopwatch with second hand
- glass stirring rod
- two 500-mL beakers
- crushed ice
- shallow tray
- thermometer
- hot plate

Chemicals
- 0.024 *M* and 0.048 *M* potassium iodate solutions
- 0.016 *M* and 0.032 *M* sodium bisulfite solutions
- 0.010 *M* copper sulfate solution
- 1% starch solution

Discussion

Several factors can influence the rate of a chemical reaction, including the presence of a catalyst, the concentrations of reactants, and the temperature at which the reaction is run. This experiment explores the effects these factors have on the rate of a reaction.

A catalyst is a substance that increases the rate of a reaction without itself being consumed. A catalyst affects the reaction rate by lowering the energy barrier (activation energy) for the reaction. You will determine the effectiveness of a catalyst by timing a chemical reaction first with and then without the catalyst present.

The rate of a chemical reaction depends on the number of times reactant particles can collide with one another. As the concentration of one or more reactants is increased, the total number of collisions possible is increased. You will study the effect of concentration on reaction rate by timing the same reaction run with various concentrations of reactants.

Temperature is a measure of the average kinetic energy of a system. Temperature affects the rate of a chemical reaction not only by influencing the number of total collisions possible but also by influencing the number of collisions that can get over the energy barrier for the reaction. You will look at this factor by timing a chemical reaction at three temperatures—room temperature, a temperature above room temperature, and a temperature below room temperature.

The chemical reaction used in this experiment is

$$2\,IO_3^- + 5\,HSO_3^- \rightarrow I_2 + 5\,SO_4^{2-} + H_2O + 3\,H^+$$

Although this is a complicated oxidation–reduction reaction, it is very easy to time. As the reaction occurs, the iodine formed reacts with starch added to the reaction mixture, forming a deep blue color. The reaction is considered complete once the solution turns blue.

Procedure

Part A Effect of Catalyst on Reaction Rate

Test 1 With the Catalyst

1. Pour 20 mL of 0.024 *M* potassium iodate solution into a 50-mL beaker, and add 8 drops of copper sulfate solution (the catalyst) to the beaker.
2. Pour 20 mL of 0.016 *M* sodium bisulfite solution into a second 50-mL beaker, and add 5 drops of starch solution to the beaker.
3. Record the start time shown on your stopwatch.
4. Start the stopwatch as you pour the contents of the first beaker into the second beaker, stirring the mixed solutions continuously. Stop the stopwatch when the reaction mixture turns deep blue.
5. Record the stop time and calculate the amount of time required for the color change to occur.

Test 2 Without the Catalyst

1. Pour 20 mL of 0.024 *M* potassium iodate solution into a 50-mL beaker.
2. Pour 20 mL of 0.016 *M* sodium bisulfite solution into a second 50-mL beaker, and add 5 drops of starch solution to the beaker.
3. Record the start time shown on the stopwatch.
4. Start the stopwatch as you pour the contents of the first beaker into the second beaker, stirring the mixed solutions continuously. Stop the stopwatch when the reaction mixture turns deep blue.
5. Record the stop time and calculate the amount of time required for the color change to occur.

Part B Effect of Concentration on Reaction Rate

Test 1

1. Pour 20 mL of 0.048 *M* potassium iodate solution into a 50-mL beaker, and add 8 drops of copper sulfate solution to the beaker.
2. Pour 20 mL of 0.016 *M* sodium bisulfite solution into a second 50-mL beaker, and add 5 drops of starch solution to the beaker.
3. Record the start time shown on the stopwatch.

4. Start the stopwatch as you pour the contents of the first beaker into the second beaker, stirring the mixed solutions continuously. Stop the stopwatch when the reaction mixture turns deep blue.
5. Record the stop time and calculate the amount of time required for the color change to occur.

Test 2

1. Pour 20 mL of 0.024 *M* potassium iodate solution into a 50-mL beaker, and add 8 drops of copper sulfate solution to the beaker.
2. Pour 20 mL of 0.032 *M* sodium bisulfite solution into a second 50-mL beaker, and add 5 drops of starch solution to the beaker.
3. Record the start time shown on your stopwatch.
4. Start the stopwatch as you pour the contents of the first beaker into the second beaker, stirring the mixed solutions continuously. Stop the stopwatch when the reaction mixture turns deep blue.
5. Record the stop time and calculate the amount of time required for the color change to occur.

Test 3

1. Pour 20 mL of 0.048 *M* potassium iodate solution into a 50-mL beaker, and add 8 drops of copper sulfate solution to the beaker.
2. Pour 20 mL of 0.032 *M* sodium bisulfite solution into a second 50-mL beaker, and add 5 drops of starch solution to the beaker.
3. Record the start time shown on your stopwatch.
4. Start the stopwatch as you pour the contents of the first beaker into the second beaker, stirring the mixed solutions continuously. Stop the stopwatch when the reaction mixture turns deep blue.
5. Record the stop time and calculate the amount of time required for the color change to occur.

Part C Effect of Temperature on Reaction Rate

Test 1 Below Room Temperature

1. Determine the ambient air temperature in the laboratory and record it in column 2, row 2, of the Part C data table.
2. Fill a 500-mL beaker about one-fourth full with crushed ice, and add about 300 mL of water. (It is fine to "eyeball" how much water you add. The beaker should be about three-fourths full after you add the water.)
3. Pour 20 mL of 0.024 *M* potassium iodate solution into a 50-mL beaker, and add 8 drops of copper sulfate to the beaker.

4. Pour 20 mL of 0.016 *M* sodium bisulfite solution into a second 50-mL beaker, and add 5 drops of starch solution to the beaker.
5. Place both beakers in the shallow tray, and carefully pour ice water into the tray until the water level in the tray is about even with the liquid level in the beakers. Leave the beakers in this ice bath until the solutions reach a temperature of about 10°C. Record the temperature of the solutions in column 2, row 3, of the Part C data table.
6. Record the start time shown on your stopwatch.
7. Start the stopwatch as you pour the contents of the first beaker into the second beaker, stirring the mixed solutions continuously. Stop the stopwatch when the reaction mixture turns deep blue.
8. Record the stop time and calculate the amount of time required for the color change to occur.

Test 2 Above Room Temperature

1. In a 500-mL beaker, heat about 300 mL of water to about 65°C.
2. Repeat steps 3 and 4 of Part C Test 1.
3. Place both beakers in the shallow tray, and carefully pour the hot water into the tray until the water level in the tray is about even with the liquid level in the beakers. Leave the beakers in this water bath until the solutions reach a temperature of about 40°C. Record the temperature of the solutions in column 2, row 4, of the Part C data table.
4. Record the start time shown on your stopwatch.
5. Start the stopwatch as you pour the contents of the first beaker into the second beaker, stirring the mixed solutions continuously. Stop the stopwatch when the reaction mixture turns deep blue.
6. Record the stop time and calculate the amount of time required for the color change to occur.

Name ______________________________

The Clock Reaction Report Sheet

Part A Effect of Catalyst on Reaction Rate

Test	Amount of $CuSO_4$ Catalyst Present (drops)	Start Time	Stop Time	Amount of Time Needed for Color Change(s)
1	8			
2	0			

Part B Effect of Concentration on Reaction Rate

Test	KIO_3 Concentration	$NaHSO_3$ Concentration	Start Time	Stop Time	Amount of Time Needed for Color Change(s)
Data from Part A, Test 1	0.024 *M*	0.016 *M*			
1	0.048 *M*	0.016 *M*			
2	0.024 *M*	0.032 *M*			
3	0.048 *M*	0.032 *M*			

Part C Effect of Temperature on Reaction Rate

Test	Temperature (°C)	Start Time	Stop Time	Amount of Time Needed for Color Change(s)
Room-Temperature Data from Part A, Test 1				
1 Below Room Temperature				
2 Above Room Temperature				

Questions

1. Doubling the number of drops of copper sulfate added should have little to no effect on the rate of reaction. Why?

2. If some of the water that was used to warm the reactants in Test 2 of Part C spilled into either of the beakers before they were mixed, would this have the effect of increasing or decreasing the rate of reaction relative to the value you reported?

3. What relationship did you find between the temperature of the reaction and its rate? Explain this relationship in terms of the collisions among reactants.

15 Upset Stomach

Objective

- To measure and compare the acid-neutralizing strengths of antacids

Materials Needed

Equipment
- two 50.00-mL burets
- buret stand with clamps
- three 250-mL Erlenmeyer flasks
- mortar and pestle
- weigh-dishes
- balance
- 10-mL pipets
- well-plates

Chemicals
- 0.50 *M* HCl solution
- 0.50 *M* NaOH solution
- phenolphthalein indicator
- various brands of antacid tablets

Discussion

Overindulging in food or drink can lead to acid indigestion, a discomforting ailment that results from the excessive secretion of hydrochloric acid, HCl, by the stomach lining. An immediate remedy is an over-the-counter antacid, which consists of a base that can neutralize stomach acid. In this experiment, you will add an antacid to a simulated upset stomach. Not all the acid will be neutralized, however, and so you will then determine the effectiveness of the antacid by determining the amount of acid that remains.

This is done by completing the neutralization with another base, sodium hydroxide, NaOH. The reaction between hydrochloric acid and sodium hydroxide is

HCl	+	NaOH	→	NaCl	+	H_2O
hydrochloric acid		sodium hydroxide		salt		water

When the concentrations of the HCl and NaOH used are the same, as is the case in this experiment, the volume of NaOH needed to neutralize the HCl remaining in the "relieved" stomach after the antacid has done its job is equal to the volume of HCl *not* neutralized by the antacid.

Procedure

Part A Preparing the Upset Stomach

1. Fill a 50.00-mL buret with 0.50 *M* HCl. Remove any air bubbles from the tip by letting a few milliliters of liquid run out the tip.
2. Record the initial volume to two decimal places, remembering from the Taking Measurements experiment how to look at the meniscus.
3. Deliver approximately 30 mL of the HCl solution to a 250-mL Erlenmeyer flask and record the final volume in the buret, again to two decimal places. This flask represents your "upset stomach." So as not to upset your eyes, you *are* wearing your safety glasses, right?
4. Add 2 drops of phenolphthalein indicator to the flask.
5. Repeat this procedure to produce additional flasks containing 30 mL of the HCl solution, so that you have one flask for each antacid you are going to test.

Part B Adding the Antacid

1. Record the brand and active ingredient of each of your antacids on the report sheet.
2. Crush and grind the first antacid tablet with a mortar and pestle. (**Note:** Alka-Seltzer need not be crushed.)
3. Carefully transfer all of the resulting fine powder of the first antacid to a weigh-dish, and then determine and record the mass of the weigh-dish plus powder. Repeat this crushing and weighing procedure for additional antacids.
4. Carefully transfer each antacid from its weigh-dish to an upset stomach flask prepared in Part A. Be sure to label each flask, then swirl each flask for a few minutes while being careful not to spill. These flasks now represent "relieved upset stomachs." (The solutions should remain colorless.)
5. Determine and record the mass of the empty weigh-dish.

Part C Completing the Neutralization (choose one of the two following procedures)

Macroscale Procedure

1. Fill the other 50.00-mL buret with 0.50 *M* NaOH. Remove air bubbles from the tip by letting a few milliliters of liquid run out.
2. Record the initial volume of NaOH in the buret to two decimal places.
3. Carefully deliver this solution, in small increments, to a "relieved upset stomach" flask. Initially you will see a pink color form and then fade away as you swirl the flask. As you get closer to the point of complete neutralization (called the end point), the pink color will persist for longer periods of time. Be sure to add the NaOH more slowly (dropwise if necessary) as you approach the end point. You should add only enough NaOH to allow a light pink color to persist for at least 1 minute. Do not overshoot the end point.
4. Record the final volume of NaOH remaining in the buret.

5. Repeat this neutralization procedure for the remaining brands of antacid.

Microscale Procedure

1. Use a pipet to transfer 20 drops of the solution in the first "relieved upset stomach" flask to each of five wells of a plastic well-plate.
2. Fill a clean pipet with 0.50 *M* NaOH solution, and then carefully and slowly add the solution *dropwise* to one of the wells. Keep a careful count of the number of drops added, and add base until a light pink color persists for at least 30 seconds.
3. Record the number of drops in the report sheet, and then repeat step 2 for the remaining wells.
4. Repeat the procedure for the remaining "relieved upset stomachs."

Name ______________________________

Upset Stomach Report Sheet

Data

Show calculations for all items marked with an asterisk (*).

Parts A and B Preparing the Upset Stomach and Adding the Antacid

	Antacid 1	Antacid 2	Antacid 3
1. Brand of antacid used	________	________	________
2. Active ingredient	________	________	________
3. Initial volume of HCl in buret	________ mL	________ mL	________ mL
4. Final volume of HCl in buret	________ mL	________ mL	________ mL
5. Volume of HCl added to Erlenmeyer flask*	________ mL	________ mL	________ mL
6. Mass of weigh-dish plus crushed antacid	________ g	________ g	________ g
7. Mass of empty weigh-dish after antacid transferred	________ g	________ g	________ g
8. Mass of antacid added to "upset stomach"*	________ g	________ g	________ g

Part C Completing the Neutralization

Macroscale Procedure

	Antacid 1	Antacid 2	Antacid 3
9. Initial volume of NaOH in buret	________ mL	________ mL	________ mL
10. Final volume of NaOH in buret	________ mL	________ mL	________ mL
11. Volume of NaOH added to complete neutralization*	________ mL	________ mL	________ mL
12. Volume of acid neutralized by NaOH (equal to volume of NaOH added)	________ mL	________ mL	________ mL
13. Volume of acid neutralized by antacid*	________ mL	________ mL	________ mL
14. Volume of acid neutralized for every gram of antacid*	________mL/g	________mL/g	________mL/g

Name ______________________

Microscale Procedure

Antacid 1	Well 1	Well 2	Well 3	Well 4	Well 5
9. Drops of "relieved upset stomach" solution	20	20	20	20	20
10. Drops of NaOH added to complete neutralization	______	______	______	______	______
11. Drops of acid neutralized by NaOH (equal to drops of NaOH added)	______	______	______	______	______
12. Drops of acid neutralized by antacid*	______	______	______	______	______

Antacid 2	Well 1	Well 2	Well 3	Well 4	Well 5
9. Drops of "relieved upset stomach" solution	20	20	20	20	20
10. Drops of NaOH added to complete neutralization	______	______	______	______	______
11. Drops of acid neutralized by NaOH (equal to drops of NaOH added)	______	______	______	______	______
12. Drops of acid neutralized by antacid*	______	______	______	______	______

Antacid 3	Well 1	Well 2	Well 3	Well 4	Well 5
9. Drops of "relieved upset stomach" solution	20	20	20	20	20
10. Drops of NaOH added to complete neutralization					
11. Drops of acid neutralized by NaOH (equal to drops of NaOH added)					
12. Drops of acid neutralized by antacid*					

Questions

1. List the antacids you tested in order of neutralizing strength, strongest first:
   strongest ________________ > ________________ > ________________ weakest

2. What would be the effect on the calculated neutralizing strength for an antacid if the following errors were made?
   You spill some of the crushed antacid as you are transferring it to the "upset stomach" flask.

   You overshoot the end point in adding NaOH to the relieved stomach fluid.

16 Mystery Powders

Conceptual Chemistry Laboratory

Objectives

- To develop an effective qualitative analysis scheme
- To identify ten common household chemicals using qualitative analysis

Materials Needed

Equipment
- clear-glass well-plates
- eyedroppers
- microspatulas
- 250-mL beaker
- thermometer
- hot plate
- test tube
- 5-mL or 10-mL graduated cylinder
- latex gloves

Test Reagents
- iodine solution
- phenolphthalein
- white vinegar
- 0.3 *M* sodium hydroxide
- 70% rubbing alcohol

Household chemicals
- cornstarch, $(C_6H_{10}O_5)_n$
- white chalk, $CaCO_3$
- plaster of Paris, $CaSO_4 \cdot H_2O$
- washing soda, Na_2CO_3
- milk of magnesia, $Mg(OH)_2$
- epsom salt, $MgSO_4 \cdot 7H_2O$
- baking soda, $NaHCO_3$
- boric acid, H_3BO_3
- table sugar, $C_{12}H_{22}O_{11}$
- table salt, NaCl

Discussion

You will be given ten vials, each containing a white powder that is a common household chemical. Your task is to identify these unknowns based on their different physical and chemical properties. First, however, you must develop a qualitative analysis scheme along the lines of the one shown in Figure 1, to show how the chemicals can be systematically identified.

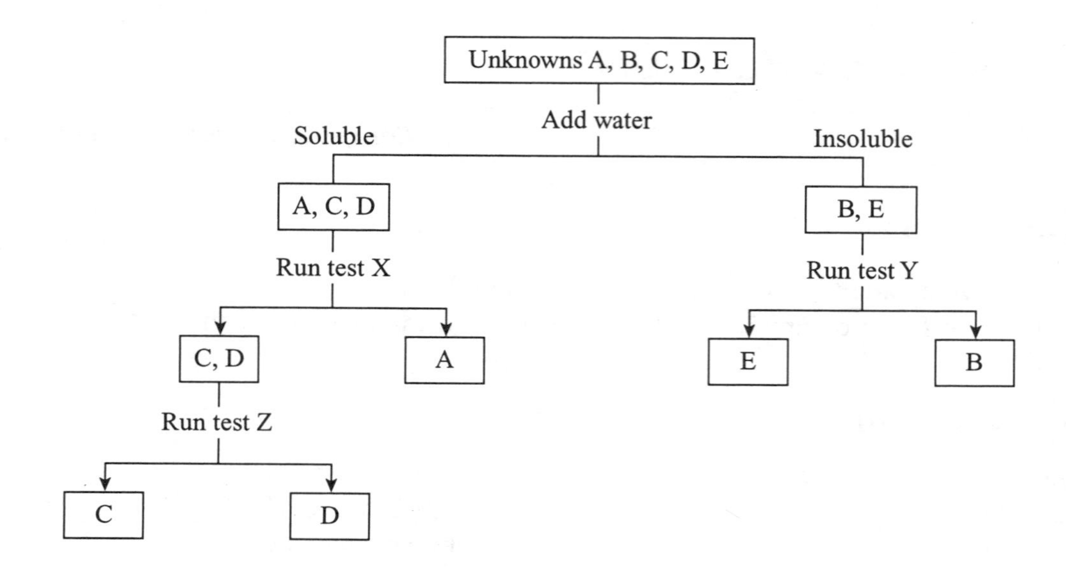

FIGURE 1 A sample qualitative analysis scheme.

Procedure

Using the tests outlined below, prepare a qualitative analysis scheme that permits you to unequivocally determine the identities of the unknowns. Start with test 1, Solubility in Water. You may then choose your own order of testing. Some orders are more efficient than others. Try to develop a scheme that minimizes the number of tests you must use. Use Table 1 as a guide to the physical and chemical properties of the household chemicals to be identified.

After getting your instructor's approval for your scheme, draw it on the report sheet, along with a data table to record your observations.

Tests

In running these tests, remember that after you add water to all your unknowns in test 1, only *one* test reagent can be added to an unknown in a well. Once you add a test reagent to an unknown on a well-plate and record your observations in your data table, rinse out the plate and start fresh. As always, you should be wearing your safety glasses.

1. **Solubility in Water** Tap a small amount of each unknown into separate wells of a well-plate. Use an eyedropper to fill each well with distilled water, and stir gently with a microspatula. (**Note:** either use a separate microspatula for each well or else rinse a single microspatula thoroughly after stirring each sample.) Record in your data table which of your unknowns are soluble in water.

2. **Iodine** Use this test on any unknown that is insoluble in water. Add a drop or two of iodine solution to each well containing an unknown as it sits undissolved in water. A deep blue color appears as the iodine complexes with cornstarch. A brownish color appears for all other unknowns insoluble in water. Record your findings in your data table.

3. **Phenolphthalein** Use this test on any unknown that is soluble in water. Add a drop of phenolphthalein to each well containing a dissolved unknown. A bright pink color appears if the solution is very alkaline, which is a positive test for milk of magnesia, $Mg(OH)_2$ (pH above 8). A medium pink color appears if the solution is only slightly alkaline, which is a positive test for washing soda, Na_2CO_3 (pH approximately 8), and a lighter pink color for baking soda, $NaHCO_3$ (pH $\geq$ 7). Record your findings.

4. **White Vinegar** Use this test on all unknowns. Add a drop or two of white vinegar to each well containing an unknown, in either a dissolved or undissolved state. The formation of bubbles is a sign of the carbonate ion, CO_3^{2-}, which decomposes to gaseous carbon dioxide upon treatment with an acid (vinegar). This test is positive for white chalk, $CaCO_3$, washing soda, Na_2CO_3, and baking soda, $NaHCO_3$. Record your findings.

5. **Sodium Hydroxide** Use this test on any unknown that is soluble in water. The test is specific for $MgSO_4$, one of the salts in epsom salt. After an unknown has been dissolved in water on a well-plate, add a few drops of O.3 *M* sodium hydroxide test reagent. Formation of a precipitate is a positive test for epson salt, $MgSO_4 \cdot 7H_2O$.

6. **Hot Water** All the water-soluble unknowns except table salt, NaCl, become markedly more soluble in warm water. This test, is therefore, specific for sodium chloride. Prepare a water bath by filling a 250-mL beaker about three-fourths full with water and heating on a hot plate to 60°C. Place several pea-sized portions of an unknown in a test tube along with 5 mL of water. Heat the test tube in the water bath held at about 60°C. No marked improvement in solubility suggests that the unknown is sodium chloride. This test takes careful observation and lots of patience.

7. **Rubbing Alcohol** Use this test on any unknown that is soluble in water. Tap a few crystals of the unknown into a well on a well-plate, and then fill the well with 70% rubbing alcohol. Boric acid, H_3BO_3, has the greatest solubility in rubbing alcohol.

TABLE 1 **Properties of Ten Household Chemicals**

Chemical	1 Solubility in Water	2 Iodine	3 Phenol-pthalein	4 Vinegar	5 Sodium Hydroxide	6 Hot Water	7 Rubbing Alcohol
Cornstarch	insoluble	positive	—	negative	—	—	—
White chalk	insoluble	negative	—	positive*	—	—	—
Plaster of Paris	insoluble	negative	—	negative	—	—	—
Washing soda	soluble	—	medium pink	positive	negative	increased solubility	insoluble
Milk of magnesia	low solubility	sometimes positive**	dark pink	negative	—	increased solubility	some solubility
Epsom salt	soluble	—	negative	negative	positive	increased solubility	insoluble
Baking soda	soluble	—	light pink	positive	negative	increased solubility	insoluble
Boric acid	moderately soluble	—	negative	negative	negative	slight increase	soluble
Table sugar	soluble	—	negative	negative	negative	increased solubility	insoluble
Table salt	soluble	—	negative	negative	negative	no effect	insoluble

* Some brands of chalk are not made of calcium carbonate and their results on the vinegar test are negative.

** Some brands of milk of magnesia contain filler starch, which will give a positive iodine test.

Name ________________________________

Mystery Powders Report Sheet

1. Write out your qualitative analysis scheme:

2. Write out your data table:

Results

If different sets of unknowns were available, specify which set you worked with:

____

Chemical	Corn-starch	White chalk	Plaster of Paris	Washing soda	Milk of magnesia	Epson salt	Baking soda	Boric acid	Table sugar	Table salt
Unknown Number										

Questions

1. Which of the tests used in this experiment measure physical properties and which measure chemical properties?

2. How many individual tests did you have to perform to identify all of the unknowns? (Review your analysis scheme or count the total number of wells and test tubes you used.) Can you spot any ways to reduce the number of tests in your scheme? If so, describe the changes you would make.

3. Is it more efficient to start with tests that split the unknowns into two groups, such as test 7, or to start with tests that are specific for single compounds, such as test 5?

17 Electrochemistry

Objectives

- To determine the relative reducing strengths of various metals
- To practice writing oxidation–reduction equations, including half reactions
- To assemble voltaic cells and predict the direction of electron flow

Materials Needed

Equipment

- steel wool
- nine medium test tubes
- wooden test tube rack with space for nine tubes
- four 5-mL graduated cylinders
- two 250-mL beakers
- flashlight bulb attached to circuit board
- two copper wire leads fitted with alligator clips
- paper towels

Chemicals

- strips of Ag, Cu, Pb, and Zn metal, 1 cm × 10 cm
- 0.10 *M* $AgNO_3$, $CuSO_4$, $Pb(NO_3)_2$, and $Zn(NO_3)_2$ solutions
- 1.0 *M* $AgNO_3$, $CuSO_4$, $Pb(NO_3)_2$, and $Zn(NO_3)_2$ solutions
- saturated NaCl solution

Discussion

Electrochemistry is the study of the relationship between electrical energy and chemical change. An electrochemical process can involve the production of an electric current via an oxidation–reduction reaction. In an oxidation–reduction reaction, one reactant loses electrons (*oxidation*) while another reactant gains electrons (*reduction*). Oxidation cannot take place without reduction, and reduction cannot take place without oxidation. Because the two always occur together, each is referred to as a half reaction. For example, solid iron metal, Fe, and dissolved nickel ions, Ni^{2+}, react to form dissolved iron ions, Fe^{2+}, and solid nickel metal, Ni. The overall reaction and the two half reactions are

Overall reaction	$Fe(s) + Ni^{2+}(aq)$	$\rightarrow$	$Ni(s) + Fe^{2+}(aq)$
Half reaction (oxidation)	$Fe(s)$	$\rightarrow$	$Fe^{2+}(aq) + 2e^-$
Half reaction (reduction)	$Ni^{2+}(aq) + 2e^-$	$\rightarrow$	$Ni(s)$

A voltaic cell is a reaction vessel designed to provide an electric current from two half reactions that take place in separate compartments. Electrons flow through a wire that connects the two compartments. To prevent a buildup of charge, it is necessary to provide a pathway, such as a salt bridge, through which dissolved ions can migrate from one compartment to the other. To gauge the strength of the electric current, a light bulb can be connected to the circuit—the greater the current, the brighter the bulb glows.

In this experiment, you will observe several oxidation–reduction reactions. Based on your observations, you'll be able to rank the relative reducing strengths (that is, the abilities to act as a reducing agent) of several metals. You can then use these reducing strengths to predict the direction of electron flow in several voltaic cells. You will verify your predictions by observing the brightness of the light bulb connected to the cells.

Procedure

Part A Relative Reducing Strengths

1. Obtain three strips each of copper (Cu), lead (Pb), and zinc (Zn).
2. Clean the surface of each with steel wool.
3. Label the test tubes 1 through 9, and arrange in the wooden rack.
4. Place the following solutions and metal strips in the tubes. Observe and record what happens in each case.

Tube 1	3 mL of 0.10 *M* $CuSO_4$ solution and one lead strip
Tube 2	3 mL of 0.10 *M* $CuSO_4$ solution and one zinc strip
Tube 3	3 mL of 0.10 *M* $Pb(NO_3)_2$ solution and one copper strip
Tube 4	3 mL of 0.10 *M* $Pb(NO_3)_2$ solution and one zinc strip
Tube 5	3 mL of 0.10 *M* $Zn(NO_3)_2$ solution and one copper strip
Tube 6	3 mL of 0.10 *M* $Zn(NO_3)_2$ solution and one lead strip
Tube 7	3 mL of 0.10 *M* $AgNO_3$ solution and one copper strip
Tube 8	3 mL of 0.10 *M* $AgNO_3$ solution and one lead strip
Tube 9	3 mL of 0.10 *M* $AgNO_3$ solution and one zinc strip

Part B Voltaic Cells

Your instructor may ask each group to construct one or more voltaic cells similar to the one shown in Figure 1. The four types of cells you may be building are specified in Table 1. Note that, even though the identities of the metals and solutions differ from one cell to the next, the basic set up is the same for all. Use 250-mL beakers. Roll up the paper towel and soak it in saturated sodium chloride solution before putting it in place as the salt bridge. Approximately 200 mL of each solution should be used. Be sure that both wires are attached securely. In the third column of the table in Part B of the report sheet, predict in which direction the electrons flow in each cell.

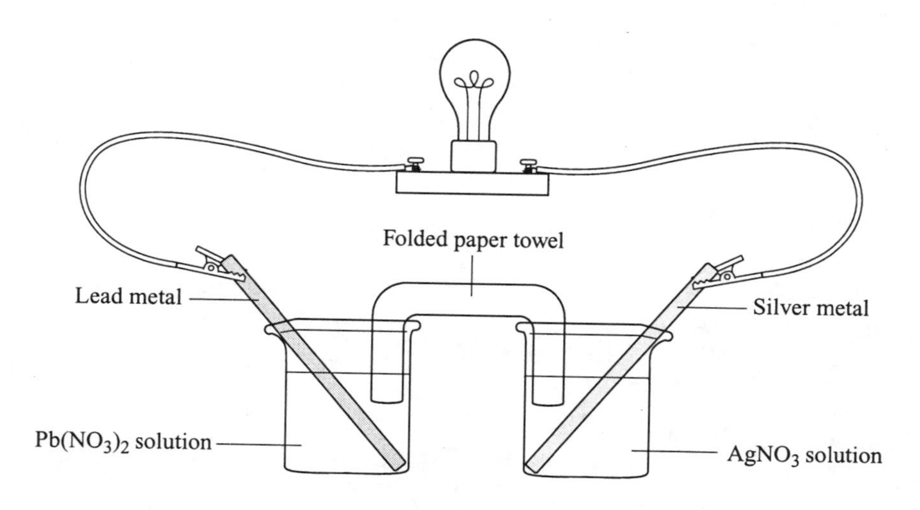

FIGURE 1

TABLE 1 Four Voltaic Cells

Cell	Beaker 1	Beaker 2
A Zn/Ag	1.0 *M* $Zn(NO_3)_2$ solution, zinc metal	1.0 *M* $AgNO_3$ solution, silver metal
B Cu/Ag	1.0 *M* $CuSO_4$ solution, copper metal	1.0 *M* $AgNO_3$ solution, silver metal
C Cu/Zn	1.0 *M* $CuSO_4$ solution, copper metal	1.0 *M* $Zn(NO_3)_2$ solution, zinc metal
D Pb/Zn	1.0 *M* $Pb(NO_3)_2$ solution, lead metal	1.0 *M* $Zn(NO_3)_2$ solution, zinc metal

Name ______________________________

Electrochemistry Report Sheet

Part A Relative Reducing Strengths

1. Record your observations:

	Copper metal	**Lead metal**	**Zinc metal**
$CuSO_4$ solution		Tube 1	Tube 2
$Pb(NO_3)_2$ solution	Tube 3		Tube 4
$Zn(NO_3)_2$ solution	Tube 5	Tube 6	
$Ag(NO_3)_2$ solution	Tube 7	Tube 8	Tube 9

2. Write the half reactions for the reduction of each metal ion:

Ion	**Half reaction**
Cu^{2+}	
Pb^{2+}	
Zn^{2+}	
Ag^{+}	

Part B Voltaic Cells

Cell	Which metal is the stronger reducing agent?	In which direction do electrons flow?	Observations	Equations
A	zinc or silver	$Ag \rightarrow Zn^{2+}$ or $Zn \rightarrow Ag^{+}$		Oxidation: Reduction: Overall:
B	copper or silver	$Ag \rightarrow Cu^{2+}$ or $Cu \rightarrow Ag^{+}$		Oxidation: Reduction: Overall:
C	copper or zinc	$Zn \rightarrow Cu^{2+}$ or $Cu \rightarrow Zn^{2+}$		Oxidation: Reduction: Overall:
D	lead or zinc	$Zn \rightarrow Pb^{2+}$ or $Pb \rightarrow Zn^{2+}$		Oxidation: Reduction: Overall:

Questions

1. Recall that a reducing agent is a substance that gives up one or more electrons to another substance, causing that substance to be reduced. The stronger the reducing agent, the greater its tendency to give up electrons. Based on your observations in Part A, rank the metals copper, lead, zinc, and silver according to their strength as reducing agents, weakest first:

weakest reducing agent ____________ < ____________ < ____________ < ____________ strongest reducing agent

Name ________________________________

2. In Part B, which voltaic cell caused the bulb to glow most brightly? Explain why the reactions in this cell cause the brightest glow.

3. Give two reasons the anode of most household disposable batteries is made of zinc rather than lead. (Recall that the anode of a battery is where oxidation occurs.)

4. Based on position in the periodic table, which would you expect to be a stronger reducing agent: lithium or zinc?

5. What advantage might there be to using lithium rather than zinc as the anode in a battery designed to run a power-hungry laptop computer?

18 Organic Molecules

Objectives

- To practice drawing representations of organic molecules
- To distinguish between structural isomers, conformers, and stereoisomers
- To become familiar with common functional groups

Materials Needed

- molecular model kit

Discussion

There are a number of ways to draw organic molecules, including full structural formulas, condensed formulas, and stick structures. For hexane, C_6H_{14}, these three representations are

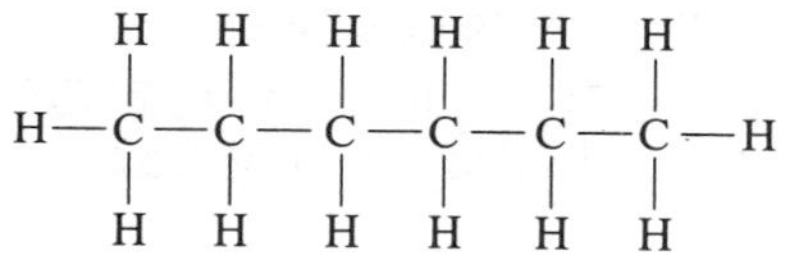

```
    H   H   H   H   H   H
    |   |   |   |   |   |
H — C — C — C — C — C — C — H
    |   |   |   |   |   |
    H   H   H   H   H   H
```

Full structural formula

$CH_3CH_2CH_2CH_2CH_2CH_3$

or

$CH_3(CH_2)_4CH_3$

Condensed formula

Stick structure

The most common way to categorize organic molecules is by the functional groups they contain. A functional group is the reactive nonalkane part of a molecule. Organic molecules containing a hydroxyl group, –OH, for example, are classified as alcohols, and those containing an amino group, $–NH_2$, are classified as amines:

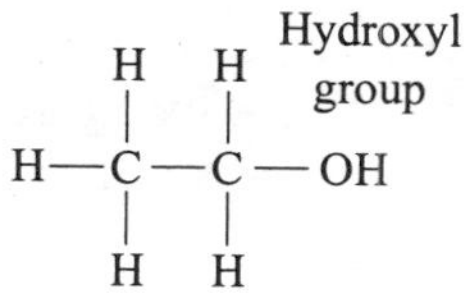

```
            Hydroxyl
    H   H   group
    |   |
H — C — C — OH
    |   |
    H   H
```

Ethanol (an alcohol)

```
            Amino
    H   H   group
    |   |
H — C — C — NH2
    |   |
    H   H
```

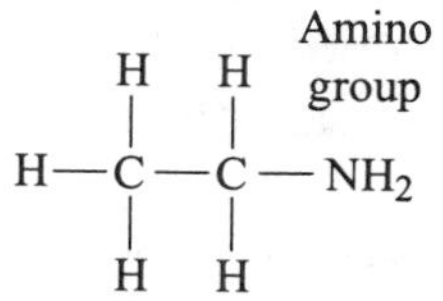

Ethylamine (an amine)

There are other relationships between organic molecules that can be helpful in working with them. In this laboratory exercise, you will examine *conformers* (different rotations of the same molecule), *structural isomers* (molecules that have the same molecular formula but their atoms connected in different order), and *stereoisomers* (molecules that differ only in the three-dimensional arrangement of their atoms and cannot be converted through rotation).

Procedure

Part A Writing Full Structural Formulas

1. Complete Table A on the report sheet.

Part B Structural Isomers and Conformers

1. Build model B1
   a. Using single bonds only, join four carbon atoms together in the most linear fashion possible. (They should form a zig-zag.)
   b. Add as many hydrogen atoms as needed, making sure each carbon atom has a total of four bonds.
2. Draw the stick structure for this molecule on the report sheet.
3. Build model B2. While holding the first two carbon atoms of the molecule you just built in place, rotate the fourth carbon 180∞ so that it ends up on the other side of the third carbon.
4. Draw the stick structure for this molecule on the report sheet.
5. Compare the two diagrams you have just drawn and fill in column 2 of Table B.
6. Build model B3. Using the four-carbon model already built, remove one hydrogen atom from the second carbon. Then pull off the fourth carbon atom and its three hydrogen atoms and attach this unit to the second carbon atom. Add a hydrogen atom to the third carbon atom so that it again has four bonds.
7. Draw the stick structure for this molecule on the report sheet.
8. Compare this diagram with the two earlier ones and complete column 3 of Table B.

Part C Structural Isomers and Stereoisomers

1. Build model C1
   a. Join two carbon atoms with a double bond.
   b. To the first carbon, attach a chlorine atom pointing up and a hydrogen atom pointing down.
   c. Do the same thing on the second carbon.
2. Draw the full structural formula for this molecule on the report sheet.
3. Build model C2
   a. Join two carbon atoms with a double bond.
   b. To the first carbon, attach a chlorine atom pointing up and a hydrogen atom pointing down.
   c. To the second carbon, attach a chlorine atom pointing down and a hydrogen atom pointing up.
4. Try to superimpose model C1 on model C2.

5. Draw the full structural formula for this molecule on the report sheet and fill in column 2 of Table C.
6. Build model C3. Using only single bonds, attach four different atoms to a carbon atom.
7. Build model C4. Using another set of the five atoms making up model C3, construct the mirror image of C3:

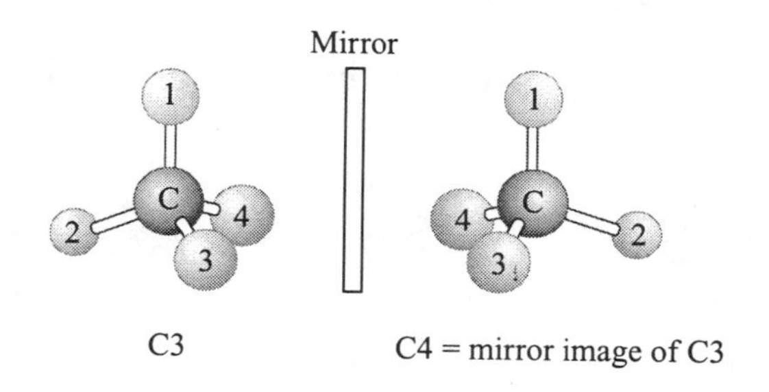

8. Try to superimpose model C3 on model C4, and then fill in column 3 of Table C.
9. Build model C5. Attach two hydrogen atoms, one chlorine atom, and one bromine atom to a carbon atom.
10. Build model C6. Using another set of C, H, Cl, and Br atoms, construct the mirror image of C5:

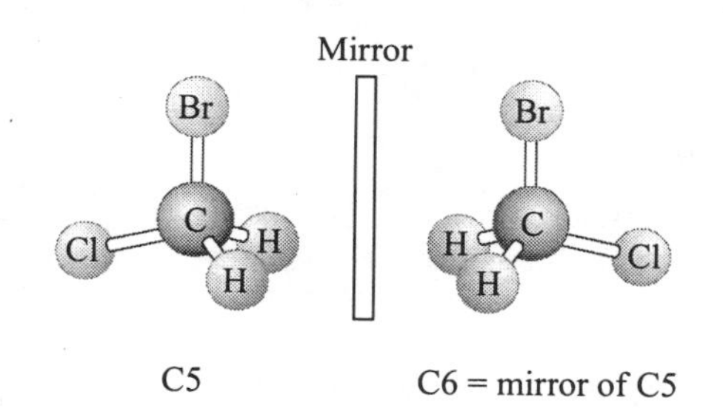

11. Try to superimpose model C5 on model C6, and then fill in column 4 of Table C.

Part D Functional Groups—Formula to Model

1. Build model D1 to represent the compound $CH_3CH_2CH_2OH$.
2. Draw a stick structure to represent model D1.
3. Build model D2 to represent the compound $CH_3CH_2OCH_3$.
4. Draw a stick structure to represent model D2.
5. Compare models D1 and D2 and complete Table D on the report sheet.

Part E Functional Groups —Model to Formula

Shown in Table E on the report sheet are six molecular models rendered in duplicate so that you can view them in three dimensions. To do this, look at each pair of images and then cross your eyes so that the two images appear to overlap. Based on what you see, complete the second and third columns of Table E. Because these models are not color-coded, you will need to deduce the identity of the atoms based on the number of bonds each atom forms. Your choices for the atoms are carbon (four bonds), nitrogen (three bonds), oxygen (two bonds), and hydrogen (one bond). Look out for multiple bonds, which are indicated in these structures by dark shading.

Part F Models of Biological Molecules

Build a model of either LSD (lysergic acid diethylamide) or morphine. Both molecules are shown in Figure 1.

LSD

Morphine

FIGURE 1

Density

19 DNA Capture

Objective

- To isolate DNA from a vegetable or fruit

Materials Needed

Equipment
- 400-mL beaker
- crushed ice
- 250-mL Erlenmeyer flask
- balance
- sharp knife
- blender
- 50-mL beaker
- stirring rod
- two 20-mL centrifuge tubes
- centrifuge
- 50-mL graduated cylinder
- glass Pasteur pipet
- microscope

Chemicals
- distilled water
- sodium chloride
- sodium bicarbonate
- additive-free liquid laundry detergent
- isopropyl alcohol, chilled
- onions, tomatoes, bananas, or other plant samples

Discussion

DNA (deoxyribose nucleic acid) makes up a small portion of the total number of biomolecules in living tissue. Because of the unique properties of DNA, it can be easily isolated from other biomolecules in certain types of tissue. The following procedure for isolating DNA takes advantage of the fact that DNA'S negatively charged phosphate groups allow it to dissolve in a solution of salt water. When relatively nonpolar isopropyl alcohol is then added to the solution, the DNA comes out of solution as a stringy goop.

Procedure

1. Prepare an ice bath by filling the 400-mL beaker about one-fourth full with crushed ice and then adding about 200 mL of water. (Tap water will do; the distilled water is for later in the procedure.)
2. In the Erlenmeyer flask, mix 1.5 grams of sodium chloride, 5.0 grams of sodium bicarbonate, 5 mL of the laundry detergent, and 120 mL of distilled water.
3. Chill the solution by placing the flask in the ice bath.
4. Obtain a vegetable or fruit from your instructor. Chop into bite-size pieces, put the pieces in the blender, and blend to a fine mush.

5. Add approximately 10 mL of the mush to the 50-mL beaker.
6. Add approximately 20 mL of the chilled salt/detergent solution and stir vigorously for at least 2 minutes.
7. If you have been given an individual container of isopropyl alcohol, place the container in the ice bath at this time.
8. Pour half of the mush mixture into one of the centrifuge tubes and the other half into the other centrifuge tube.
9. Centrifuge for about 1 minute. If all the solids have not settled to the bottom by this time, centrifuge a second time.
10. Carefully decant the solution from both tubes into a clean single 50-mL graduated cylinder. The solution should be relatively clear at this point. This solution contains DNA from your fruit or vegetable.
11. With the Pasteur pipet, slowly and gently add 10 mL of chilled isopropyl alcohol to the DNA solution so that the alcohol forms a top layer. This can be done by allowing the drops of alcohol to run down the side of the graduated cylinder.
12. Use the tip of the Pasteur pipet to gently stir the DNA solution just below the alcohol layer, gently "lifting" the solution toward the alcohol layer.
13. Look for a murkiness, and within that murkiness look for the formation of DNA strands, which will adhere to the pipet as they are brought up into the alcohol layer. (The DNA is not soluble in the alcohol.)
14. Using a pipet, transfer some of the strands to a shallow glass dish. Look at these strands under a microscope, preferably a stereoscopic microscope.

Appendix

A.1 Significant Figures Are Used To Show Which Numbers Have Experimental Meaning

Two kinds of numbers are used in science—those that are *counted or defined* and those that are *measured.* There is a great difference between a counted or defined number and a measured number. The exact value of a counted or defined number can be stated, but the exact value of a measured number can never be stated.

You can count the number of chairs in your classroom, the number of fingers on your hand, or the number of quarters in your pocket with absolute certainty. Defined numbers, which are about exact relationships and are defined as being true, are also known with absolute certainty. The defined number of centimeters in a meter, the defined number of seconds in an hour, and the defined number of sides on a square are examples.

Every measured number, however, no matter how carefully measured, has some degree of *uncertainty.* This uncertainty (or margin of error) in a measurement can be illustrated by the two meter sticks shown in Figure 1. Both sticks are being used to measure the length of a table. Assuming that the zero end of each meter stick has been carefully and accurately positioned at the left end of the table, how long is the table?

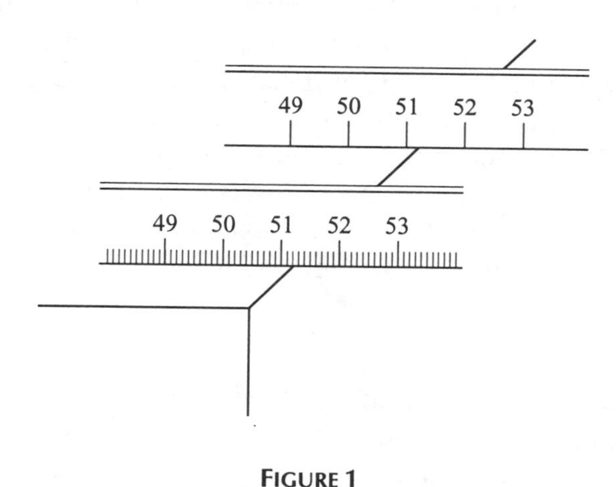

FIGURE 1

The upper meter stick has a scale marked off in centimeter intervals. Using this scale, you can say with certainty that the length is between 51 and 52 centimeters. You can say further that it is closer to 51 centimeters than to 52 centimeters; you can even estimate it to be 51.2 centimeters.

The scale on the lower meter stick has more subdivisions—and therefore greater precision—because it is marked off in millimeters. With this scale, you can say that the length is definitely between 51.2 and 51.3 centimeters, and you can estimate it to be 51.25 centimeters.

Note how both readings contain some digits that are *known* plus one digit (the last one) that is *estimated.* Note also that the uncertainty in the reading from the lower meter stick is less than the uncertainty in the reading from the upper meter stick. The lower meter stick can give a reading to the hundredths place, but the upper one can give a reading only to the tenths place. The lower one is more *precise* than the upper one. So, in any measured number, the digits tell us the *magnitude* of the measurement and the location of the decimal point tells us the *precision* of the measurement. (Figure 2 illustrates the distinction between *precision* and *accuracy*.)

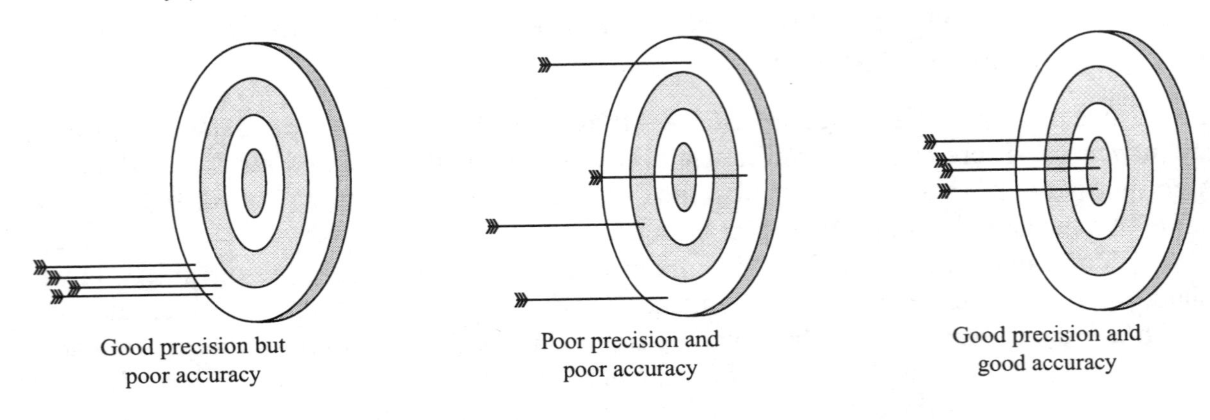

FIGURE 2

Significant figures are the digits in any measurement that are known with certainty plus one final digit that is estimated and hence uncertain. These are the digits that reflect the precision of the instrument used to generate the number. They are the digits that have experimental meaning. The measurement 51.2 centimeters made with the upper meter stick in Figure 1, for example, has three significant figures, and the measurement 51.25 centimeters made with the lower meter stick has four significant figures. The rightmost digit is always an estimated digit, and only one estimated digit is ever recorded for a measurement. It would be incorrect to report 51.253 centimeters as the length measured with the lower meter stick. This five-significant-figure value has two estimated digits (the final 5 and 3) and is incorrect because it indicates a *precision* greater than the meter stick can obtain.

Here are some standard rules for writing and using significant figures.

Rule 1

In numbers that do not contain zeros, all the digits are significant:

4.1327	five significant figures
5.14	three significant figures
369	three significant figures

Rule 2

All zeros between significant digits are significant:

8.052	four significant figures
7059	four significant figures
306	three significant figures

Rule 3

Zeros to the left of the first nonzero digit serve only to fix the position of the decimal point and are not significant:

0.0068	two significant figures
0.0427	three significant figures
0.0003506	four significant figures

Rule 4

In a number that contains digits to the right of the decimal point, zeros to the right of the last nonzero digit are significant:

53.0	three significant figures
53.00	four significant figures
0.00200	three significant figures
0.70050	five significant figures

Rule 5

In a number that has no decimal point and that ends in one or more zeros, the zeros that end the number are not significant:

3600	two significant figures
290	two significant figures
5,000,000	one significant figure
10	one significant figure
6050	three significant figures

Rule 6

When a number is expressed in scientific notation, all digits in the coefficient are taken to be significant:

4.6×10^{-5}	two significant figures
4.60×10^{-5}	three significant figures
4.600×10^{-5}	four significant figures
2×10^{-5}	one significant figures
3.0×10^{-5}	two significant figures
4.00×10^{-5}	three significant figures

A.2 There Are Rules for Rounding Off Significant Figures

Calculators often display eight or more digits. How do you round off such a display to, say, three significant figures? Three rules govern the process of deleting unwanted (nonsignificant) digits from a calculator number.

Rule 1

If the first digit to the right of the last significant figure is less than 5, that digit and all the digits that follow it are dropped:

51.234	51.2	rounded off to three significant figures

Rule 2

If the first digit to the right of the last significant figure is greater than 5 or if it is 5 followed by something other than zero, that digit and all the digits that follow it are dropped and the last retained digit is increased by 1:

51.38	51.4	rounded off to three significant figures
51.359	51.4	rounded off to three significant figures
51.3598	51.4	rounded off to three significant figures

Rule 3

If the first digit to the right of the last significant figure is 5 not followed by any other digit or if it is 5 followed only by zeros, an odd–even rule is applied If the last retained digit is even, its value is not changed and the 5 and any zeros that follow are dropped. If the last retained digit is odd, its value is increased by 1. The intention of this odd–even rule is to average the effects of rounding off:

74.85	74.8	rounded off to three significant figures
74.75	74.8	rounded off to three significant figures
74.2500	74.2	rounded off to three significant figures
89.3500	89.4	rounded off to three significant figures

A.3 Significant Figures Must Be Maintained Throughout a Calculation

Suppose you measure the mass of a small wooden block to be 2 grams and then find that its volume is 3 cubic centimeters by dipping it beneath the surface of water in a graduated cylinder. The density of the wood is the mass of the block divided by its volume. If you divide 2 by 3 on your calculator, the reading on the display is 0.6666666. However, it would be incorrect to report the density as 0.6666666 gram per cubic centimeter because to do so would be claiming a higher degree of precision than is warranted. Your answer should be rounded off to a sensible number of significant figures.

The number of significant figures allowable in a calculated result depends on the number of significant figures in the measured data and on the type of mathematical operation(s) used in calculating. There is one set of rules for multiplication and division and another set for addition and subtraction.

Multiplication and Division

The answer to a multiplication or division calculation must have the number of significant figures found in the multiplied or divided number that has the fewest significant figures. For our density example, the density value must be rounded off to one significant figure, 0.7 gram per cubic centimeter. If the mass were measured to be 2.0 grams, but the volume were still taken to be 3 cubic centimeters, the answer must still be rounded off to 0.7 gram per cubic centimeter. If the mass were measured to be 2.0 grams and the volume 3.0 or 3.00 cubic centimeters, the answer must be rounded off to two significant figures: 0.67 gram per cubic centimeter.

As you study the following examples, assume that the numbers being multiplied or divided are measured numbers.

Example A

$8.536 \times 0.47 = 4.01192$ (calculator answer)

The input having the fewest significant figures is 0.47, which has two significant figures. Therefore the calculator answer must be rounded off to 4.0.

Example B

$3840 \div 285.3 = 13.459516$ (calculator answer)

The input having the fewest significant figures is 3840, which has three significa... figures. Therefore the calculator answer must be rounded off to 13.5.

Example C

$36.00 \div 3.000 = 12$ (calculator answer)

Both inputs contain four significant figures. Therefore the correct answer must also contain four significant figures, and the calculator answer must be written as 12.00. In this case, the calculator gave too few significant figures.

Addition and Subtraction

The answer to an addition or subtraction should not have any digits beyond the last digit position common to all the numbers being added or subtracted.

Example A

 34.6
 18.8
+15
 68.4 (calculator answer)

The last digit position common to all numbers is the tens place. Therefore the calculator answer must be rounded off to 68.

Example B

20.02
20.002
+20.0002
60.0222 (calculator answer)

The last digit position common to all numbers is the hundredths place. Therefore the calculator answer must be rounded off to 60.02.

Example C

345.56
–245.5
100.06 (calculator answer)

The last digit position common to both numbers is the tenths place. Therefore the calculator answer must be rounded off to 100.1.